IN THE LIGHT

DON ALEXANDER

In the Light   Don Alexander

DEDICATED TO:

An understanding of the Light of life.

PROLOGUE

The curse has devoured the earth and those who dwell therein are desolate: therefore, the inhabitants of the earth are burned and few men left.

Isaiah, the son of Amoz   B.C. 760

Holy Bible

I beheld the earth and it was without form and the heavens had no light.
I looked and there was no man and all the birds of the heavens were gone.

Jeremiah, the son of Hilkiah   B.C. 629

Holy Bible

CHAPTER   ONE

Those who spend their short lives
studying astrology, astronomy and astrophysics
tell us that the sun represents 99.86% of the
mass of our solar system. The planets, comets,
asteroids, and miscellaneous interstellar masses
make up the other 0.14%.

The sun's diameter is estimated at
1,392,000 kilometers and its total mass is
330,000 times the mass of Earth. The sun orbits
the Milky Way at a distance roughly equal to
25,000 light years with a velocity approaching
370 kilometers per second completing one orbit

each 250,000,000 years. The chemical composition of the sun is 75% hydrogen, 23.31% helium and 1.69% oxygen, carbon, neon, and iron plus trace elements heavier than helium. This 1.69% of the sun's mass is 5,628 times the mass of Earth. The sun is just one of the 200 billion or more stars within Milky Way.

The Milky Way is one of an estimated 200 billion galaxies and its velocity is calculated at 550 to 600 kilometers per second. Milky Way's diameter is approximately 120,000 light years (one light years is just under ten trillion kilometers – the distance light travels in one year at a velocity of around 300,000 kilometers per second).

The number of individual stars within

In the Light   Don Alexander

the known universe number into the trillions. The sun converts its hydrogen mass through nuclear fusion into helium at the rate of 620 million metric tons per second. The fused helium contains less mass than the converted hydrogen. The excess mass resulting from the fusion of hydrogen into helium radiates out from the sun in the form of pulsating waves of electromagnetic energy which we call sunlight.

Light can be described as the absence of darkness. The sun is not the only source of light since electromagnetic energy waves are created by chemical reactions other than the conversion of hydrogen into helium. The electromagnetic light waves contain individual packets of light energy called photons. Photons are created when sub-atomic particles of matter called electrons undergo temperature changes in connection with chemical processes involving heat energy.

The most common sources of light other than sunlight are the creation of waves of electromagnetic energy created during the combustion of flammable gases; the burning of molecular compounds such as wood, coal and petroleum derivatives; and the heating of various metals such as the filament in light bulbs.

A more definitive understanding of light is achieved by considering the micro universe in the form of the sub-atomic particles which make up both the micro and macro masses including both inorganic matter and living organisms. All matter within the universe is composed of what scientists call elements. The known elements total 112 of which 91 predate the appearance in space of our solar system.

The other 21 elements resulted from transmutation of some of the original elements through the process of radioactive decay, nuclear fission or nuclear fusion. A significant number of the elements are found in all organic and inorganic masses and are essential to the very existence of the physical structure of bacteria, plants, insects, animals and humans.

All the elements are composed of microscopic divisions of matter which scientists call atoms. The atoms of each element are composed of protons, neutrons and electrons.

In the Light   Don Alexander

The only difference between atoms of different elements is the number of protons within the atom's nucleus. For example, a hydrogen atom contains one proton whereas an atom of uranium contains 92 protons.

Hydrogen is the lightest element and uranium is one of the heaviest elements. The nucleus of an atom is simply protons and neutrons in close proximity to each other. The nucleus of each atom of each element is surrounded by one or more orbiting electrons.

Protons are positively charged, electrons are negatively charged, and neutrons are electrically neutral. The term "electrical" with respect to atoms refers to nothing more than the attracting and repelling forces holding atoms together.

Atomic forces binding atoms together are called "the strong nuclear force and the weak nuclear force." Both forces are known to exist but the origin thereof is an unsolved mystery which baffles theoretical physicists striving to understand the relationship between energy, matter and momentum.

There are three primary sources of energy in the universe referred to as nuclear energy, radiation energy and magnetic energy. Energy in motion is kinetic energy. Kinetic energy at rest is potential energy, and kinetic energy produces both heat and pressure. Within the atom of an individual element, kinetic energy drives the orbiting electrons.

In the Light   Don Alexander

A charged atomic particle in motion (such as electrons) generates magnetic energy. Nuclear energy (the strong nuclear force and the weak nuclear force) keeps the nucleus of every atom (except hydrogen) from self-destructing due to the repelling force exerted by positively charged protons within the nucleus (hydrogen atoms contain a single proton).

Radiation energy pulses in waves when an element changes into another element (such as hydrogen into helium) through nuclear fusion within the core of stars. Each star functions as a giant nuclear reactor. There are billions of galaxies and an individual galaxy contains billions of stars.

The process of nuclear fusion within

stars emits continuous electromagnetic radiation (light) and nuclear energy resulting from the uncoupling of the strong and weak nuclear forces bonding the nucleus of individual atoms within the fusing elements.

The nuclear energy released in the form of waves of radiation creates massive magnetic energy waves which push against the radiation waves at right angles. This joint energy force is what causes the spin of orbiting masses in space.

The trillions of stars in space are emitting continuous waves of nuclear energy and magnetic energy waves which means that space is not empty between galaxies. The continuous pulsing of energy waves from each galaxy is what causes the expanding universe over a visible radius (through the Hubble telescope) of roughly thirteen billion light years.

The velocity of energy waves is reinforced by the continuous energy waves pulsing behind them causing the galaxies to eventually recede from each other faster than the speed of light.

CHAPTER   TWO

A proton in a hydrogen atom is identical to a proton within an atom of uranium. All neutrons are also the same but atoms of a particular element may contain more or less neutrons than another atom of the same element and are referred to as isotopes of the element.

For example, the isotope of uranium (U-235) is much rarer than the common uranium atom (U-238) but is more susceptible to transmutation through nuclear fission. Hence, U-235 is the fuel originally used in nuclear power plants.

In the Light   Don Alexander

   Atoms that are electrically neutral have
the same number of electrons orbiting the
nucleus as the number of protons within the
nucleus. All of the atom outside its nucleus is
mostly empty space. Electrons within this space
orbit the nucleus completing billions of cycles
every millionth of a second. The amazing speed
of electron orbit makes the atom appear to be a
solid mass.

   Each proton has a mass equal to the mass
of 1,836 electrons. The mass of an electron is
0.000000000000000000000000009 of one
gram which is very close to zero mass. It takes
1,839 electrons to equal one neutron's mass.

   The protons and neutrons are confined
within the tiny nucleus of the atom and are in

constant motion.

The "strong nuclear force" keeps the nucleus from flying apart by overcoming the mutual repulsion of the positively charged protons.

To illustrate the reality of how tiny the nucleus of an atom is, if an atom had a diameter of five miles, the nucleus would be about the size of a man's fist.

Atoms can gain, lose or share electrons during chemical reactions with atoms of another element. An atom that loses one or more electrons becomes a positive ion whereas an atom that gains one or more electrons becomes a negative ion.

A negative or positive ion results when

the negative charge on the atom's total electrons does not match the positive charge on the atom's total protons.

If an atom easily gives up electrons, its valence is positive, and atoms that tend to gain electrons have a negative valence.

Sodium tends to lose its one electron and thus has a valence of (+1). Chlorine tends to accept one electron from another atom and therefore has a valence of (-1). Negative ions can chemically bond with positive ions.

Thus, a molecule of ordinary table salt consists of one atom of sodium linked to one atom of chlorine. This type of chemical interaction between the atoms of the known elements is how all the matter in the universe, both organic and inorganic, is structured.

The nucleus makes up nearly all the mass of an atom. Protons and neutrons which make up the nucleus are roughly 100,000 times smaller than the atom. Electrons are not known to be composed of smaller particles of matter whereas protons and neutrons are composed of smaller particles called quarks.

Each proton and each neutron is made up of three quarks. Quarks can be manipulated by researchers within a science laboratory to form other particles of matter besides protons and neutrons but such particles are highly unstable and break down within a tiny fraction of a second. Therefore, these unstable particles are not found outside the laboratory.

In the Light   Don Alexander

Each electron has inherent energy in proportion to its orbiting velocity. The incredible velocity of orbiting electrons resembles vibration rather than orbit. The strong nuclear force binding protons within the atom's nucleus also appears to vibrate and is believed to be the actual source of gravitational attraction between masses.

The positively charged protons within the nucleus exert a force on orbiting negatively charged electrons that keeps them within the atom when the atom is not involved in a chemical reaction, nuclear fission, or nuclear fusion. The inherent energy within an electron generates resistance to the attracting force of the nucleus. The more energy the electron has, the farther from the nucleus it will be.

Consequently, electrons are arranged in energy shells at varying distances from the nucleus as determined by the level of their inherent energy. Electrons with the least energy are located in the inner shells and those with higher energy levels are in the outer shells.

Each electron energy shell is identified by a number or letter. The shell closest to the nucleus is shell #1 or shell K. The other shells, in order of increasing distance from the nucleus, are numbered 2 through 7 or labeled L through Q. Each shell can hold a limited number of electrons. Shell 1 can hold no more than 2 electrons. Shell 2 can hold 8 electrons, shell 3 can hold 18, shell 4 can hold 32, shell 5 can hold

50, shell 6 can hold 72, and shell 7 can hold 98.

However, the outer shells are never completely filled. The number of filled shells is determined by the number of electrons contained within the atom. An atom that has lost all its electrons will become a positively charged free nucleus.

There can also be free electrons (negative charge), and free neutrons (neutral charge) as the result of radioactive decay, nuclear fission and nuclear fusion. In the atoms of radioactive elements the nucleus will change as the atom gives off radioactive particles.

The change in the nucleus may be rearrangement of its protons and neutrons or the actual loss of one or more. If only the arrangement of the nucleus changes, gamma rays are emitted from the atom. If the number of protons changes, alpha or beta radiation is given off. When an atom loses one or more protons, it changes to an atom of a different element.

If one or more neutrons escape from the nucleus, the atom becomes an isotope of the radiating element. All elements heavier than bismuth are radioactive as well as the isotopes of some of the lighter elements. Isotopes of nearly all the elements can be created by bombarding their atoms with subatomic particles.

The atomic number denotes how many protons the atom of an element contains, and the mass number identifies the sum of the protons and neutrons within the nucleus. Atomic weight is the weight of an atom expressed in "atomic mass units" (amu). One amu or "dalton" equals 1/12 the weight of an atom of carbon 12. There are 602 billion trillion amu in one gram.

All atoms of the same element have the same number of protons. Since every hydrogen atom contains one proton, the atomic number of hydrogen is 1. The atomic numbers range successively up to 94 for plutonium because this element has 94 protons in each atom. Elements with more than 94 protons in each of its atoms can be created by scientists in the laboratory.

There exists more than one isotope for most of the elements. For example, hydrogen has three. The most common has no neutron in the nucleus of each atom. In the other two isotopes, the nucleus contains one to two neutrons. The mass number is used to distinguish the three isotopes; hydrogen 1, hydrogen 2 and hydrogen 3. These isotopes are also called protium, deuterium and tritium respectively.

Most of the lighter elements contain about the same number of protons and neutrons in the nucleus of their atoms. The heavier elements have more neutrons than protons. The heaviest elements have about three neutrons for every two protons. U-238, for example, has 92 protons and 146 neutrons.

Atoms of different elements which have the same mass number but different atomic

numbers are called isobars. The isobars argon
and calcium have a mass number of 40 but
argon's atomic number is 18 (18 protons) and
calcium's atomic number is 20 (20 protons).

The way an atom of an element behaves
during a chemical reaction is largely determined
by the number of electrons in its outermost
electron shell. When atoms combine and form
molecules, electrons in the outermost shell
are either transferred from one atom to another
or shared between atoms.

The number of electrons involved in the
chemical reaction is referred to as "valence."
The atoms of some elements can have more than
one valence depending on the number and kind
of atoms they can combine with.

In the Light   Don Alexander

Electrons are restricted to a limited set of motions, each of which has a specific energy value. These motions are referred to as quantum states or energy levels. When an electron is in a given quantum state, it does not give off or absorb energy. An atom can lose or gain energy only when one or more electrons change from one quantum state to another.

Electrons seek the lowest state of energy but only one electron at a time can occupy each quantum state. When the lower states are filled, other electrons are forced to occupy higher states. When all electrons are in the lowest available state, the atom is in "ground state" which is the normal condition for atoms at ordinary temperatures.

In the Light   Don Alexander

When matter is heated to a few hundred degrees, sufficient energy is then available to raise one or more electrons to a higher energy level. The atom is then transformed into an "excited state" which lasts for a fraction of a second. An excited electron quickly drops to a lower state and continues dropping until the atom returns to its ground state.

During each succeeding drop, the electron gives off a tiny packet of radiant energy called a photon. The energy of the photon equals the difference between the two energy levels the electron passed through. These photons are detected as visible light and other forms of electromagnetic radiation.

One neutron and one proton can occupy

each quantum state in the nucleus of an atom. A light nucleus has about the same number of protons and neutrons but a proton and neutron in the same state do not have the same amount of energy because each proton is electrically repelled by all other protons in the nucleus thereby increasing the energy of each proton.

In a heavy nucleus, the difference in energy levels between protons and neutrons is significant and more low energy states are available for neutrons than for protons. This helps explain why a heavy nucleus contains more neutrons than protons.

Most of the 91 elements found in and on Earth (as contrasted to the elements created in scientific laboratories and nuclear reactors) are in compound form. They are combined with other elements forming soil, rocks, gas, liquids, minerals, crystals, etc. Oxygen and silicon are the most plentiful elements in Earth's crust and make up 3/4 of the crust's weight. A few elements are found in pure form in small amounts such as gold, copper, carbon and sulfur.

It is easy to confuse energy, force and power. Energy is the word used to describe the ability to make things happen, like raising the temperature of liquids, gases and solids. Energy propels, directs and accelerates all types of matter; produces light; binds subatomic particles within the nucleus of all atoms; and numerous other activities classified as "work."

The amount of activities which can be

accomplished depends upon the strength of the force used and the distance through which it moves. Power measures the rate at which the work is performed.

All matter is held together by the energy which prevents the nucleus of every atom from self-destructing. Therefore, energy existed within the universe before any of the elements came into being. An atom of any of the elements is mostly empty space invaded by energy emitting from the tiny nucleus and the orbiting electrons.

The incredible amount of energy present within the strong nuclear force which offsets the repelling force of the positively charged protons is measured by Einstein's formula: energy equals mass times the speed of light squared (E = MC squared).

The destructive power of nuclear weapons results from releasing the strong nuclear force from within the atoms of certain radioactive elements.

Since energy preceded the formation of matter and matter is composed of atoms, atoms were perfectly designed so that energy could hold them together. In other words, atoms had to come into existence within the universe in the form of their existing irreducible level of complexity.

Although closely intertwined, energy and matter are not the same. It is obvious from the very structure of matter that it is mostly energy in motion. Yet, the force which holds matter together is not matter; it is energy. Energy and matter were applied to a specific preplanned design and living beings were introduced into a physical environment that had already been carefully crafted to nurture and sustain them.

The questions that physicists, chemists, astronomers, and biologists wrestle with are extremely basic but not simple. How did the universe come into existence? Was the birth of the universe accidental or designed? When did the universe appear? Is the universe progressing from a beginning to an end? Which elements make up the bulk of the matter contained in the universe? How did this combination of elements originate? How were living organisms introduced into the universe? What forces account for the delicate balance between energy, matter, and momentum? Where did viruses and bacteria come from? Where and how did the hierarchy of bacteria, insects, plants, animals and humans originate? How do inorganic compounds acquire life? What exactly is life? Is there life after physical death for humans? For other living organisms? How and when does the human fetus acquire life?

CHAPTER   THREE

The known elements, including the
nuclear elements resulting from nuclear fission
and nuclear fusion, are: hydrogen, beryllium,
sodium, magnesium, lithium, potassium,
calcium, scandium, titanium, vanadium,
chromium, manganese, iron, cobalt, nickel,
copper, zinc, boron, carbon, nitrogen, oxygen,
fluorine, neon, aluminum, silicon, phosphorus,
sulfur, chlorine, argon, gallium, germanium,
arsenic, selenium, bromine, krypton, rubidium,
strontium, yttrium, zirconium, niobium,
molybdenum, technetium, ruthenium, rhodium,

In the Light   Don Alexander

palladium, silver, cadmium, indium, tin,

antimony, tellurium, iodine, xenon, caesium,

barium, hafnium, tantalum, tungsten, rhenium,

osmium, iridium, platinum, gold, mercury,

thallium, lead, bismuth, polonium, astatine,

radon, francium, radium, rutherfordum,

dubnium, seaborgium, bothrium, hassium,

meitnreium, ununhexium, damstadium,

roentgerium, copemicium, ununtrium,

ununquadum, ununpentium, ununsepium,

ununoctium, lanthanum, actinium, cerium,

thorium, protactinium, praseodymium,

neodymium, uranium, promethium, neptunium,

samarium, plutonium, europium, americium,

gadolinium, curium, terbium, berkelium,

dysprosium, californium, holmium, einsteinium,

erbium, fermium, thulium, mendelevium, ytterbium, nobelium, lutetium, and lawrencium.

The physical molecular structure comprising the bodies of all life forms ranging from a single-celled bacterium to humans consists of the primordial elements (non-nuclear elements). The hundred trillion or so single cells the human body is composed of contain nothing but oxygen, carbon, hydrogen, nitrogen, calcium, phosphorus, potassium, sulfur, chlorine, sodium, magnesium, iron, cobalt, copper, zinc, iodine, selenium and fluorine, plus a few trace elements.

Life itself is by far the strangest phenomenon of all. It does not consist of the primordial elements nor any of the non-primordial elements and therefore could never have originated from any energy, matter or momentum within the entire universe.

Therefore, life is not part of the physical body and thus is not dependent upon a physical manifestation. Yet, the body cannot move, feed nor reproduce without life.

Upon death the physical body reverts back to the elements from which it originated. Where life returns upon separation from the body is thus far incomprehensible to proponents of the "big bang theory" coupled with the "theory of evolution."

Big bang advocates postulate that the primordial universe sprang from nothingness in the initial form of a tiny speck of "extremely condensed" and "very heavy pure energy." The temperature of this speck of pure energy was so great that "energy flowed into matter and back into energy" until this tiny speck of primordial energy exploded in shell burst fashion (or just expanded) into all the energy and matter contained in the known universe.

At the exact moment of the big bang, the tiny speck of pure energy was flowing between energy and matter such that billions of trillions of cubic miles of hydrogen, helium and primordial debris were flung across a vacuum of nothingness. This spewed out mass cooled over 385,000 years thereby allowing the purely random formation of all the other elements [total of 118 as listed in the Periodic Table of the

Elements] without design, purpose or order.

The "other elements" were created accidentally out of the hydrogen, helium and primordial interstellar debris created by the big bang (or the big eternally expanding energy bubble).

Evolutionists further assume that this big bang (or fantastic expansion of a tiny dot of primordial hot flowing energy/mass) produced plasma, solids, liquids and gases; antimatter (the opposite of matter); dark energy (invisible energy); and dark matter (invisible matter).

The big bang spewed forth a tiny fraction more matter than antimatter. For every billion particles of antimatter there were a billion and one particles of matter. Antimatter canceled out an equal amount of matter through nuclear collisions.

The tiny speck of pure energy/matter blew up or simply expelled energy and matter with so much force that the excess matter not canceled by antimatter became everything in the universe that we see today (galaxies, stars, planets, suns, moons, meteorites, asteroids, comets, bacteria, viruses, insects, plants, animals, humans, etc.).

This big bang happened in a "planck" moment of time (rough estimation = one thousand billion trillionth of one clock second).

The pure energy unit which blew up or simply expanded was trillions of degrees hot and moving extremely fast (faster than the speed of light) so that in a few planck units of time the universe expanded several billion light years in diameter (one light year equals 5.88 trillion miles).

Then, after 200 million more years of slow cooling, the other non-gaseous elements randomly formed within billions of stars which randomly formed from vast clouds of hydrogen.

These stars were basically billions of gigantic balls of hydrogen gas created by the big bang and held together by the force of gravity being exerted throughout the primordial universe by dark matter and dark energy.

Dark matter and dark energy randomly appeared from nothingness in just the right randomly formed proportions (by the big bang)

to accidentally impose order out of random
and chaotic events transpiring faster than a
couple planck units of time.

Such randomly directed primordial
energy forced the universe to continue
expanding in an orderly and totally balanced
fashion. This randomly created but precise
order and balance permitted a galaxy named
"Milky Way" (a grouping of billions of
individual stars, moons, planets, comets,
asteroids, and other interstellar debris) to
accidentally manipulate itself into the precise
orbit and exact distribution of inherent
gravitational forces to keep Planet Earth safely
in perpetual orbit around the star nearest Earth
(our sun).

In the Light   Don Alexander

The big bang randomly fixed Earth's orbit at just the right distance, the right speed, orbital tilt, and surface temperature to permit the eventual evolution of various life forms.

The galaxies are receding from each other at a high velocity indicating that the universe is expanding rapidly and there are cosmic background microwave radiation echoes emitting from deep space (as detected from Earth).

Big bang advocates proclaim that the continuously expanding universe and the microwave echoes are sufficient to prove the big bang theory to be fact because the big bang obviously triggered the expanding universe and the microwave echoes are due to the heat left over from the big bang (heat which originated over fourteen billion years ago). Billions of galaxies and trillions of stars (one of which is Earth's sun) are not generating any deep space heat according to this rationale.

The average temperature of the universe is calculated by big bang advocates to be about 450 degrees below zero, Fahrenheit (less than three degrees above absolute zero, Kevin).

Evolutionists and big bang advocates also point to Einstein's "theory of relativity" as proof of the big bang. Einstein postulated that the energy inherent in matter is equal to the mass of the matter (expressed as weight in kilograms) times the speed of light squared. Applying this formula to the tiny speck of pure energy/matter is awkward because the mass of the tiny speck expressed in kilograms is pure speculation supported only by imagination.

At the exact planck second during which the tiny speck of pure energy was flowing into the same tiny speck of matter, the big bang occurred and spewed out billions of trillions of cubic miles of matter containing inherent energy which later accidentally formed the elements.

The tiny speck of pure energy/matter is described as an atom sized dot to be consistent with the claim that a single singularity equal to approximately nothingness triggered the big bang.

The tiny dot of pure energy/matter was left over after a previous universe collapsed upon itself sort of like a giant star running out of hydrogen fuel and collapsing into "a black hole." This primordial black hole contained energy and matter so dense that its inherent gravitational field sucked in all the energy and matter that ventured across its "event horizon."

Captured energy and matter continued to plummet toward "the singularity" (extremely condensed energy and matter) which sucked so hard that not even light could escape. The black hole emitted "Hawking radiation" (named after Stephen Hawking, a theoretical physicist) until it disappeared into "a parallel universe."

Even though light cannot escape a black hole, this tiny speck (actually an unknowable singularity) vomited out trillions of cubic miles of matter and energy. And, fortunately for humans, this random vomiting event happened before Hawking radiation could empty the black hole causing it to disappear into a parallel universe.

Nine billion years after the big bang, the debris created cooled enough to permit the formation of gigantic gaseous masses orbiting in interstellar space. Some of the randomly formed hydrogen, peppered with interstellar debris, randomly organized itself and formed the solar system containing the planet Earth.

CHAPTER FOUR

The universe just keeps getting bigger and bigger but retains all the matter and energy not canceled by antimatter during the big bang such that today the universe is probably about 150 billion light years across (150 billion times 5.88 trillion miles).

Dark energy is now pulling the universe apart so that about 100 billion years into the future the universe will collapse in on itself like letting the air out of a balloon.

There will probably be another big bang creating a whole new universe. This process of expanding and shrinking has possibly repeated itself numerous times in the undated past permitting a steady progression of new universes to be created out of the initial primordial state such that the big bang is probably a repetitive event.

Evolutionists and big bang advocates proclaim that all "true scientists" have accepted the random chance formation of the universe and spontaneous generation of life forms as scientific fact. However, there are many "true scientists" who never accepted the accidental formation of everything from nothingness and who also fathered every existing field of science between the first and twentieth centuries:

Leonardo da Vinci, Johann Kepler,

In the Light   Don Alexander

Francis Bacon, Blaise Pascal, Robert Boyle,

John Ray, Nicolaus Steno, Thomas Burnet,

Athanasius Kircher, John Wilkens, Walter

Charleton, Sir William Petty, Isaac Barrow,

Increase Mather, Nehemiah Grew, Galileo,

Robert Hooke, William Harvey, Christian

Huygens, Tycho Brahe, Nicholas Copernicus,

Isaac Newton, William Whiston, John

Woodward, Carolus Linnaeus, Jonathan

Edwards, William Herschel, John Harris,

Gottfried Wilhelm Leibnitz, John Flamsteed,

William Derham, Cotton Mather, John

Hutchinson, Gustavus Brander, Jean Deluc,

Richard Kirwan, James Parkinson, Michael

Faraday, Humphrey Davy, George Cuvier,

Timothy Dwight, Benjamin Silliman, Charles

In the Light   Don Alexander

Bell, William Buckland, Charles Babbage,
David Brewster, John Herschel, John Dalton,
William Kirby, Jedidiah Morse, Benjamin
Barton, Samuel Miller, John Kidd, Peter Mark
Roget, Thomas Chalmers, William Prout,
Samuel F. B. Morse, Joseph Henry, Matthew
Maury, James Simpson, James Joule, Adam
Sedgwick, William Whewell, Henry Rogers,
Louis Agassiz, James Dana, John William
Dawson, George Stokes, Charles Piazzi,
Rudolph Virchow, Phillip H. Gosse, Gregor
Mendel, Louis Pasteur, Henri Fabre, Lord
William Thomas Kelvin, Joseph Lister, Joseph
Clerk Maxwell, Bernhard Riemann, John Bell
Pettigrew, George Romanes, Richard Owen,
Edward Hitchcock, Sir Henry Rawlinson, Sir

In the Light   Don Alexander

Joseph Henry Gilbert, Thomas Anderson, Sir
William Huggins, Balfour Stewart, P. G. Tait,
John Murray, James Glaisher, Edward H.
Maunder, William Mitchell Ramsey, Lord John
W. S. Rayleigh, Alexander MacAlister, A. H.
Sayce, John Ambrose Fleming, Howard A.
Kelly, George Washington Carver, Charles
Stine, Douglas Dewar, Paul Lemoine, William
Ramsey, Wernher von Braun, Sir William
Abney, A. Rendal Short, L. Merson Davies, and
Sir Cecil P. G. Wakeley.

The big bang theory assumes that a
nebula (huge cloud of hydrogen containing
interstellar debris) erupted from the primordial
black hole. This nebula contained hydrogen,
helium, oxygen, iron, magnesium, aluminum,

carbon dioxide, methane, sodium, and modest quantities of the rest of the elements. It randomly formed into a big spiraling cloud which included dust-sized pieces of rocks and metals.

The nebula collapsed in a couple of eons due to its own weight and inherent gravitational forces and contracted. It began spinning and flattened into an enormous disc. Internal particles collided, stuck together and formed planetesimals.

Some of the planetesimals randomly combined to form nine planets. Others formed moons, asteroids, meteorites, and comets. The remaining nebula debris randomly formed a modest sized star (our sun). The newly formed nine planets with their moons, along with dust rings, asteroids, meteorites, comets and miscellaneous debris orbited the newly formed sun.

Since most living creatures must have water in one form or another, big bang advocates explain that Earth formed initially as gas and molten debris from its core to its surface and thus was void of water and had no atmosphere. An interstellar mass smacked into Earth and blew out a chunk of matter which by gravitational attraction locked into orbit around Earth becoming Earth's moon.

Earth and its moon cooled for eons of time during which they were pelted by meteorites, comets, asteroids and miscellaneous interstellar debris; some of which contained water and water vapor randomly formed by chance from hydrogen and oxygen gases.

Earth, being a ball of molten rock and gases, instantly vaporized any inherent water and, as Earth cooled somewhat, water vapor on Earth's surface was scoured away by the solar wind (electromagnetic particles with a velocity in excess of a million miles per hour radiating out from the sun and extending far beyond Earth).

Consequently, only when Earth cooled sufficiently to permit random formation of atmospheric gases and water vapor, could water laden impacting masses from interstellar space create Earth's streams, rivers, lakes and oceans.

In the Light   Don Alexander

After gaining an atmosphere, Earth had to endure billions (or perhaps trillions) of impacts with comets, meteorites, and asteroids small enough not to disrupt Earth's orbiting mass and yet large enough to retain ice or water while being superheated by Earth's atmosphere.

The water in liquid or solid form (carried to Earth by such "larger masses") survived the tremendous heat generated upon impact with Earth and deposited their inherent water and/or ice on or inside Earth.

In the field of theoretical physics (unlimited personal imagination) built upon quantum mechanics (combining proven and unproven theories in the field of physics and theoretical mathematics), there are three additional popular evolutionary explanations as to how the universe came into being: the "string theory;" the "M theory;" and "the theory of everything." However, these concepts are in the hypothetical stages and remain mostly imagination and conjecture aimed at explaining away impossibilities within the big bang and evolutionary explanations of how everything sprang from nothingness.

Such purely speculative, unfounded hypotheses ignore every proven and established law of physics and are demonstrated by "a wise old scientist" chalking incomprehensible (even to a venerable physicist) mathematical equations on a blackboard and finally exclaiming: "eureka!!! There is no God!!!"

In the Light   Don Alexander

It is worth remembering that
mathematical equations are meaningless when
the scales of such equations are balanced with
nothing but imagination and unfounded
speculation. A valid scientific equation must be
based upon proven facts.

Another ruse used to make singularities
appear more plausible is to invent new words
and phrases which describe imaginary physical
forces. Such forces are represented by
mathematical equation models jammed with
unknowable values pertaining to purely
theoretical elementary particles such as the
"Higgs boson" and the "gravitron." Such
unproven and thus imaginary sub-atomic
structures are just two of several theoretical

particles of atomic matter believed to be
partially responsible for various discernible
manifestations of energy bound up within the
atoms of the elements.

Every form of organic or inorganic
matter, other than pure elements, consists of
chemical compounds which under heat and
pressure break down into molecules which
further break down into atoms of the known
elements, and then into subatomic particles
comprising the atoms.

This simplistic description of molecular
structure is actually the present state of scientific
knowledge concerning the composition of the
universe and reveals absolutely nothing
concerning the origin of energy, matter and

momentum.

More than 2,500 scientists, engineers, and researchers have now gone public stating that the big bang is both delusional and intellectually insulting.

Recently, Stephen Hawking admitted that he has been wrong for thirty years about black holes and the loss of information into parallel universes. Intellectually honest theoretical physicists can no longer ignore the fact that Einstein, Planck, Hawking and other theoretical physicists were simply wrong.

Their mathematical models were based on equations using faulty mathematical assumptions. They also borrowed formulas from other physicists that are based on hypotheses that have never been elevated to the status of a theory and therefore cannot be honestly presented as scientific fact.

Many of these false hypotheses are being presented in our public schools as proven facts in spite of the known fact that such hypotheses have never been elevated to a theory which can be scientifically tested and cannot ever become a fact if scientific testing of the theory has not yielded anticipated and consistent results.

Einstein's hypothesis concerning general relativity was accepted as fact based upon one experiment of deflected starlight during a total eclipse of our sun. Thus, big bang, evolution, black holes, parallel universes, and other theoretical hypotheses based only on personal imagination are still merely hypotheses and not theories and certainly not facts.

Einstein's theory of general relativity is based on faulty mathematics (the use of a constant for what is now proven to be a variable on one side of the equation $E = MC$ squared) as recently demonstrated by feedback from orbiting satellites and the Hubbel Space Telescope).

The two "basic proofs" quoted by big bang supporters are "the expanding universe" and the "microwave radiation echoes emitting from deep space;" which are both true but do not in any way whatsoever support the imaginary big bang. The microwave radiation echoes are related to electromagnetic radiation emitting from every galaxy. The expanding universe is driven by a combination of galactic emissions and magnetic fields created by the intersecting forces of radiation and magnetic waves pulsing from every galaxy.

It is now observed that the speed of light is not the maximum attainable speed with respect to a real mass with straight line velocity. The Hubble Space Telescope has detected galaxies receding from other galaxies faster than the speed of light.

Newtonian gravitational concepts in terms of description (quantitative) have been proven reliable over three centuries. However, Newton never discovered nor tried to explain the actual source (qualitative) of gravitational

forces.   It now appears that gravity is a force
that results from the continuous vibration of the
"strong nuclear force" that binds the nucleus of
every atom in the universe thereby binding all
matter together with a force in proportion to the
mass of the matter and inversely proportional to
the distance between masses.

CHAPTER   FIVE

When considering the various "educated explanations" as to the origin of the universe and its life forms, the elementary definitions of imagination, opinion, hypothesis, theory, law of cause and effect, and established facts can be relied upon to separate truth from fiction.

An opinion may be based on imagination or knowledge accumulated through observations concerning the foundation for the opinion or a combination thereof. Opinions generally contain some facts mixed with bias.

A hypothesis is a conclusion based upon

one or more observations colored by personal opinion and unsupported by scientific experiments which consistently yield the same result.

A theory is a hypothesis which is subject to being tested by repetitious and valid scientific experiments. An untested theory can never rise to the level of fact.

An established fact is given birth by a theory which has withstood repetitive scientific testing yielding the same results conforming to the "law of cause and effect." A "happening," "state of being," or "event" is the effect; and the factor which gives birth to the effect is the cause.

The law of cause and effect states: any

factor in whose presence the effect fails to occur cannot be the cause; and conversely, any factor in whose absence the effect occurs cannot be the cause.

The big bang theory coupled with the evolution of all life forms descending from a single-celled living organism is a hypothesis and not a theory because there have been no scientific tests which produced repetitive and predictable results. The law of cause and effect cannot verify as factual an untested theory which is based only on biased and random observations mixed with imagination pertaining to a singularity (an event that has never been verified).

The explanation for a singularity which

violates the laws of physics (scientifically established facts) is highly unlikely and proclaiming such explanation to be factual is the height of intellectual dishonesty.

From the beginning of human history upon Planet Earth, scientifically oriented individuals have pondered the origin of the sun, moon, stars, and Earth's life forms. The concept of spontaneous generation of primordial life forms and evolution of humans from lower life forms was proposed more than a thousand years before the birth of Charles Darwin along with the assumption that the celestial bodies are eternal without beginning or end.

Darwin was a botanist and knew nothing about molecular structure or the genetic reproduction of living organisms. Charles Lyell who was a friend of Darwin was an atheistic lawyer who wanted to discredit the Bible. Lyell proposed the "geological column" which has been proven a fiction by petrified trees and whale fossils extending through multiple sedimentary layers supposedly deposited over millions of years apart.

Another approach to dating Earth is to find some ongoing process such as radioactive decay within the atoms of radio-active elements; make some very favorable conclusions concerning equilibrium between radioactive and non-radioactive elements; apply mathematical formulas to such equilibrium relationship; and derive the age of Earth therefrom. Two such measurement techniques massaged by disciples of Darwin are carbon dating and radiometric dating.

Carbon dating is a highly controversial and inconsistent dating technique. The method is based on the rate of decay of the radioactive carbon isotope, carbon-14, which is formed in the upper atmosphere through the effect of cosmic ray neutrons upon nitrogen-14.

The carbon-14 is rapidly oxidized and enters Earth's organic life through photo-synthesis (plants) and the food chain (animals). Carbon-14 also enters the earth's oceans in an atmospheric exchange and dissolved carbonate. Plants and animals, which utilize carbon in organic functions absorb carbon-14 during their lifetimes.

The totally false assumption is that the earthbound carbon exists in equilibrium with the carbon-14 in the atmosphere; which means that the number of carbon-14 atoms and non-radioactive carbon atoms stays approximately the same over time. As soon as a plant or animal dies, it ceases its carbon intake.

Thereafter, there is no replenishment of radioactive carbon-14, only decay. The carbon-14 dating mathematical model is scientifically invalid because the atmospheric equilibrium between carbon-14 and non-radioactive carbon does not exist; and there is no evidence whatsoever that such equilibrium ever existed.

Carbon dating advocates have resorted to

In the Light   Don Alexander

Dendrochronology (tree ring dating) to create a smoke screen to draw attention away from the "equilibrium dilemma." They claim that Dendrochronology allows them to determine past concentration levels of Carbon-14 in the atmosphere by measuring the Carbon-14 to Carbon-12 ratios in tree rings.

The unavoidable errors inherent in this red herring cross reference is that no trees have been shown to exceed 4,500 years in age. The Methuselah Tree in southern California has been designated as the oldest living tree, and it has been dated at roughly 4,500 years.

Carbon-14 advocates use tree rings from dead trees thought to perhaps overlap the Methuselah Tree to mathematically determine ages exceeding 4,500 years. They determine whether a dead tree's age exceeds the ancient Methuselah Tree's age by ring patterns, and then they assume that the dead trees are older through a comparison of ring patterns, carbon ratios, etc.

It has repeatedly been demonstrated that dead tree ring patterns are typically inconsistent; and living trees can show dissimilar patterns caused by differing soil nutrients, direction of prevailing sunlight, fire history, distance to water sources, etc.

Radiometric Dating is another yardstick employed by evolutionists to determine the age of Earth. Radiometric dating techniques are predicated upon the natural decay of radioisotopes. An isotope is one or more atoms which have the same number of protons in their nuclei, but a different number of neutrons.

Radioisotopes are unstable isotopes. They spontaneously decay emitting radiation in the process thereby making them radioactive. They continue to decay going through various transitional states until they finally reach stability.

For example, Uranium-238 (U238) is a radioisotope. It will spontaneously decay until it transitions into lead-206 (Pb206). The numbers 238 and 206 represent the atomic mass for U238 and Pb206. The Uranium-238 radioisotope goes through 13 transitional stages before stabilizing into Lead-206: (U238> Th234> Pa234> U234>

Th230> Ra226> Po218> Pb214> Bi214>
Po214> Pb210> Bi210> Po210> Pb206).

In this instance, Uranium-238 is called the "parent" and Lead-206 is called the "daughter." By measuring how long it takes for an unstable element to decay into a stable element, and by measuring how much daughter element has been produced by the parent element within a specific rock sample, devout evolutionists believe they are able to determine the age of the rock.

This belief is based upon totally unreasonable assumptions: (1) no daughter elements were originally present in the rock from which the sample was extracted; (2) the rate of radioactive decay is an unwavering constant; and (3) no contamination of any kind has occurred within the rock strata (leeching).

The diffusion rate of helium gas from within zircon crystals buried deep within Earth's basement granite demonstrates the gross errors buried within the radiometric dating model.

The amount of helium gas remaining within the zircon crystals (as verified by controlled laboratory testing) date the zircons at approximately six thousand years as opposed to the billion and a half years established by radiometric dating.

There are some additional things to consider when pondering the possible age of Earth as well as irresponsible age estimates. "Preponderance of the evidence" is a legal concept pertaining to logical considerations utilized by judges and juries in reaching a verdict.

The evidence placed before the court by the litigants may be all circumstantial or partly circumstantial and partly direct in content. Fingerprints, DNA matching, ballistics, eye witness testimony, and medical records are examples of direct evidence whereas motive, opportunity, lack of alibi, and behavior pattern are examples of circumstantial evidence.

In civil cases, when the overwhelming majority of the evidence before the court points in one direction, the verdict is usually rendered accordingly.

Evidence with respect to Earth's age will

fall mainly into the circumstantial category because the formation of the planet preceded the arrival of living organisms.

Earth, in geologic terms, from the perspective of evolutionists is quite old (a few billion years). On the other side of the age issue are creationists who maintain the planet is fairly young (a few thousand years). Obviously, both estimates cannot be valid. What does the planet's magnetic field reveal concerning the age question?

Both sides agree that the magnetic field deflects much of the cosmic radiation that would otherwise destroy life on Earth. Scientists around the world have taken exact measure-ments of this magnetic field beginning in 1829 and continuing through the present date. These precise, ongoing measurements show exponential deterioration following a predictable curve.

Since 1829, there has been a seven percent deterioration in Earth's magnetic field. When the curve of fixed rate deterioration is graphed over time, it becomes quite apparent that roughly 22,000 years ago Earth's magnetic field would have been as strong as the sun's magnetic field; and that around 10,000 A.D.

Earth's magnetic field will be too weak to keep cosmic radiation from destroying all life on the planet. Life on Earth would have been impossible prior to around 20,000 B.C. and cease to exist by 10,000 A.D. or sooner.

CHAPTER   SIX

Earth's axial speed is deteriorating (slowing down) and scientists measure this slowdown in "leap seconds." Every eighteen months an additional second is required for Earth to complete one axial rotation.

If Earth's spin is slowing, logic dictates that at some time in the past, Earth was spinning faster than it is today. Earth's rate of spin is a significant factor with respect to life on the planet because axial velocity directly contributes to the "Coriolis Effect" which in turn affects the circulation of Earth's atmosphere and weather patterns.

Today's scientists acknowledge that the moon is gradually drifting away from Earth. Here again, at some past date the moon was much closer to Earth than it is today. Based on the current rate of moon-drift, 1.2 billion years ago, the moon would have been touching Earth.

In accordance with the "Inverse Square Law" pertaining to the attraction between Earth and its moon, if the moon was one third closer to Earth than it is today the gravitational effect on our tides would be nine times greater causing massive flooding in each hemisphere each tidal cycle.

Since the oceans cover about 75% of Earth's surface, such frequent and unrestrained flooding each tidal cycle would prevent air breathing organisms from evolving into humans.

Earth's oceans contain measurable quantities of Aluminum, Antimony, Barium, Bicarbonate, Bismuth, Calcium, Carbonates, Chlorine, Chromium, Cobalt, Copper, Gold, Iron, Lead, Lithium, Manganese, Magnesium, Mercury, Molybdenum, Nickel, Potassium, Rubidium, Silicon, Silver, Sodium, Strontium, Sulfate, Thorium, Tin, Titanium, Tungsten, Uranium and Zinc.

Rivers and other waters draining into the oceans dump chemical solids at a measurable rate. Comparison between amounts already in the oceans with the rates at which more are being dumped indicates a fairly young Earth.

In the Light   Don Alexander

The level of current sediments would have been deposited within a few thousand years.

Comets lose rather than gain matter so that over time comets deteriorate to extinction. Short period comets with known, predictable orbits around Earth (the orbital period) should have completely deteriorated within 10,000 to 12,000 years. Yet, the short period comets (like Haley's comet) are still orbiting Earth thereby indicating a young planet.

Jupiter is five times farther from the sun than Earth and is losing heat twice as fast as it gains heat from the sun. Jupiter's surface is still hot indicating a young solar system. In addition, Jupiter's moon, Ganymede, has a strong magnetic field which indicates it is still hot.

In the Light   Don Alexander

Saturn's rings are drifting away from the planet
and should have cleared Saturn if the planet is
billions of years old.

In the first century A.D., Earth's human
population was approximately 250 million, and
passed one billion in 1800; three billion in 1962,
five billion in 1985, and six billion in 1999.

Considering all the factors incorporated
into the "doubling time cycle" (amount of time
required to double Earth's human population)
such as technological knowledge, plagues, wars,
famines, etc., Earth's current human population
would be achieved starting with one male and
one female roughly sixty centuries ago.

The actual age of the universe and the
stars, planets, asteroids, meteorites, comets, and

interstellar debris provides not a single clue as to the question of origin. Regardless of whether the Earth is several billion years old or several thousand years of age, longevity provides zero information as to origin.

On the other hand, the actual age of the universe and Planet Earth is critical to every evolutionist because evolution of bacteria into humans requires eons of time according to the basic theory of evolution.

The Biblical record of the creation of "the heavens and the earth" does not indicate the elapsed time between the origin of the heavens and the earth and the time at which the earth was "without form and void, covered by water and existing in complete darkness."

Consequently, the actual age of both the universe and Planet Earth is not critical to Biblical veracity.

It is currently acknowledged by unbiased scientists and some world famous evolutionists that Earth's fossil record does not support evolutionary theories as indicated by the following quotes:

"When it comes to the origin of life there are only two possibilities: creation or spontaneous generation. There is no third way. Spontaneous generation was disproved one hundred years ago, but that leads us to only one other conclusion, that of supernatural creation. We cannot accept that on philosophical grounds; therefore, we choose to believe the impossible:

that life arose spontaneously by chance." ("The Origin of Life," Scientific American, 191, P. 48, May, 1954)

"In the years after Darwin, his advocates hoped to find predictable progressions. In general, these have not been found -- yet the optimism has died hard, and some pure fantasy has crept into the textbooks." ("Evolution and the Fossil Record," Science,   vol. 213, July, 1981, pg. 289)

"Evolution is unproved and unprovable. We believe it because the only alternative is special creation, and that is unthinkable (" Sir Arthur Kieth, renowned British evolutionist).

"The pathetic thing is that we have scientists who are trying to prove evolution which no scientist can ever prove." (Nobel prize winning physicist Robert A. Millikan)

"The theory of evolution is one of the strangest phenomena of humanity; it is entirely destitute of proof." (World famous geologist from Canada, Sir William Dawson)

"The Darwinian theory of descent has not a single fact to confirm it in the realm of nature. It is not the result of scientific research, but purely the product of imagination.(Professor Fleischmann, University of Erlangen zoologist}

"There is not the slightest evidence that any of the major [animal] groups arose from any other." (Dr. Austin H. Clark, world famous American biologist)

"Darwin's theory of natural selection has

never had any proof..." (Dr. Richard
Goldschmidt, Professor of zoology, University
of California)

"The Darwinian approach has consistently been to find some supporting fossil
evidence, claim it as proof for evolution," and
then ignore all the difficulties. It is, in fact, a
common fantasy..." (Roger Lewin, science
journalist)

Dr. Steven Jay Gould of Harvard
University and Niles Eldredge (curator of the
American Museum of Natural History), both
evolutionists and atheists, confessed their
awareness that there is a systematic absence of
transitional forms or "missing links" in the fossil
record. Darwin expected scientists to find

literally billions of missing links. But, where are such fossils? According to Eldredge and Gould, the fossil record is totally void of the much anticipated transitional forms.

Regardless of the age of the universe and Planet Earth, it is axiomatic that the elements composing everything in the universe as well as the physical structure of all living organisms existed before the known universe or any "collapsed universes" or "parallel universes" could possibly have appeared in the darkness of space.

Moreover, regardless of whether the elements emerged accidentally from nothingness or were designed and then created by a creative power existing outside of space and time, we do know today the elemental makeup of the known universe as well as Earth's crust, surface, oceans, and atmosphere.

Earth's core contains solid and liquid metals but the percentage of each is unknown. The metals are believed to be nickel, iron, gold, platinum, cobalt, tin and tantalum.

Earth's crust, surface, oceans and atmosphere consist entirely of oxygen (49.2%), silicon (25.7%), aluminum (7.50%), iron (4.71%), calcium (3.39%), sodium (2.63%), potassium (2.40%), magnesium (1.93%), hydrogen (0.87%), titanium (0.58%), chlorine (0.19%), phosphorous (0.11%), manganese (0.09%), carbon (0.08%), sulfur (0.06%), barium (0.04%), nitrogen (.04%), fluorine

(0.03%), and trace elements (totaling 0.49%).

The entire universe is composed of hydrogen (90.71%), helium (8.59%), carbon (0.02%), nitrogen (0.04%), oxygen (0.06%), plus trace elements (all the other elements totaling 0.58%).

The human body per 70 kilograms of body weight is composed of oxygen (45.5 kg), carbon (12.6 kg), hydrogen (7.0 kg), nitrogen (2.1 kg), calcium (1.0 kg), phosphorus (0.70 kg), magnesium (0.35 kg), potassium (0.24 kg), sulfur (0.18 kg), sodium (0.10 kg), chlorine (0.10 kg), iron (0.003 kg), zinc (0.002 kg),   plus trace elements present in less than one milligram quantity for each element (arsenic, chromium, cobalt, copper, fluorine, iodine, manganese, molybdenum, nickel, selenium, silicon, and vanadium).

Considering that all energy and matter in the entire universe is composed of the elements, what is life and how did life come into being (plant life, viruses, bacteria, insects, animals, and humans)? And, are there any life forms in the universe superior to humans?

CHAPTER SEVEN

Scientists describe life as having the ability to move, feed and reproduce. By such definition, a virus is not a living organism because a virus does not feed on any nutrients nor reproduce. The metabolism of the "host cell" provides the nutrients because the host cell's DNA replicates the virus.

Plants, bacteria, insects, animals and humans have life because they move, feed and reproduce. But, the ability to move, feed and reproduce is merely the evidence of possessing life. What exactly is life? It does not consist of any elements and therefore does not consist of anything found in the universe. Thus, life is completely separate from the physical chemistry of the discernible structure making up bacteria, insects, plants, animals and humans.

One explanation is that plant, bacterium, insect, animal and human life just appeared from nothingness and then replicated by random chance without logic, design, or purpose.

The only other possible explanation is that a creator existing outside of time and space designed the physical structure of all living organisms and then imparted life which is passed on from one generation to the next generation through whatever creative process introduced the original life forms.

In the Light   Don Alexander

Only life can reproduce life by scientific definition. Neither male sperm nor female eggs possess life. What males and females can replicate is the physical structure of their offspring.

The physical structure of the body of each life form is replicated through genetic processes including DNA coding. But, neither genetic processes nor the mingling of non-living (inorganic) structures (produced through the chemical combining of the elements) can possibly initiate life.

Consequently, we can only choose to believe some impossible explanation for life (such as the big bang and spontaneous generation of bacteria which evolved into humans over eons of time); or choose to believe in an intelligent and all-powerful creator existing outside of time and space. There are simply no other rational choices.

During the course of human history, in each generation, individual humans have personally chosen to believe in spontaneous generation of life forms or to believe in an all-powerful creator.

Spontaneous generation of any life form within the universe has been proven to be scientifically as well as mathematically impossible. Nevertheless, those humans who reject, for whatever personal reason, the existence of a creator existing outside of time and space, choose to believe that everything accidentally sprang out of nothingness without design, order or purpose even though they know that such a concept is absolutely impossible.

In order to cope with a personal belief in an impossible concept, many humans have decided that they are living in an imaginary universe and that nothing is real. Others have decided that some sort of creator does exist but is unapproachable and thus unknowable.

Others have soothed their need to commune with a creator by worshiping idols they create from clay, wood, stone and metal; or choose to worship the sun, moon, stars, etc. Some choose to worship themselves or lower life forms.

It is not uncommon for humans to claim

that they have made no decision regarding the existence of a creator. There is no state of being void of a conscious decision. Before any word can be spoken or any physical act undertaken, the relevant thought must first take shape in the brain.

Therefore, humans think, say and act based on what they have chosen to believe. Thoughts, words and acts that are contrary to personal beliefs result in sorrow, grief, anxiety, depression, and sometimes, a change in what the individual chooses to believe.

A personal belief system which incorporates an all-powerful creator existing outside of time and space is usually referred to as a "religion." Over five billion humans alive today have adopted a specific religion. Most of the adherents of any religion adopt whatever belief system is taught to them as a child.

Once a religion is adopted, it is rare that the individual turns to another religion. It is more common for an individual to discard a religious belief and turn to atheism.

Approximately one third of the world's population profess to be Christians and believe in the deity of Jesus Christ. The sacred text for Christianity is the sixty-six divisions of the Holy Bible written by forty different authors between 1491 B.C. and 96 A.D.

The most astounding fact supporting the absolute veracity of the Holy Bible is that forty different humans over a period spanning nearly sixteen centuries wrote in total harmony with each other concerning the creation of the universe and all its life forms. They also described the same creator existing outside of time and space and the eternal relationship between the creator and humans.

In addition, Biblical prophets predicted hundreds of specifically detailed events spanning twenty-four centuries which came to pass exactly as predicted. Biblical authors referred to the all-powerful creator as Jehovah, Yahweh, Adonai, and I AM THAT I AM.

The English translation of the Holy Bible refers to the same single all-powerful deity as Lord, Lord God, Lord of Hosts, Creator, Almighty God, Jesus Christ, Lamb of God, Son of God, Son of Man, Holy Spirit, and Holy Ghost.

The Holy Bible declares repeatedly that the single all-powerful deity reveals Himself to humanity in both spirit and human form as God, the Father; God, the Son; and God, the Holy Spirit.

From the perspective of humans on Planet Earth pertaining to the hierarchy of intelligent life forms, there exists only demons and angels between humans and the Triune Godhead. Humans have heard God's audible voice and seen Him in both angelic (Angel of the Lord) and human form (Jesus Christ).

Angels are ministering spirits in a visible body who administer God's heavenly kingdom and are sent on specific missions to Earth to minister to humans. Demons are fallen angels who rebelled against the Godhead and were banished to various areas of the universe including Earth.

The father of rebelling angels is Lucifer who is referred to in Biblical writings as Satan, Dragon, Serpent, and Devil. Evil Spirits are spawned by Satan and his demonic host.

The Holy Bible teaches that God Almighty created the first male and female human bodies from the dust of Earth and breathed His own breath into human nostrils to give life to the created bodies. The creation of humans followed creation of the universe including celestial bodies (sun, stars, moons,

Earth, etc.), plant life and all life forms on
Earth between humans and bacteria.

Between the creation of Earth and the
creation of humans, Earth and whatever living
creatures inhabited the planet underwent God's
judgment for some unrevealed transgression
wherein Earth was left without form and void,
covered completely with water, and orbiting in
total darkness.

God designed the human body to be
immortal and self-healing such that every body
cell renews itself on a repetitive basis. God
placed His created humans in a garden paradise
and gave humans total dominion over Planet
Earth. God named the man Adam, and Adam
named his wife Eve. Like the angels, Adam and

Eve were created with a free will and could choose to rebel against God.

Adam and Eve chose to join Satan's rebellion thereby turning over dominion of Earth to Satan. They were sentenced by God to physical death through the aging process. They were cast out of the garden to toil endlessly for their food until the death sentence materialized.

Ten generations after Adam and Eve, Earth was filled with human evil, violence and total debauchery with a single exception – the family of a man named Noah. To save humanity from total self-destruction, God instructed Noah to build a huge ark to same himself, his family, and a male and female from among all non-human, air-breathing living creatures. God sent

an Earth-cleansing flood and then repopulated
Earth from Noah's family and the other living
creatures protected by the ark.

A few generations after Noah, Earth was
again filled with violence and evil. Humans
gathered in what is now modern Iraq between
the Tigris and Euphrates rivers. They conceived
and endeavored to build a city and tower
reaching into what they believed to be heaven.

God confounded human genetics and
language to scatter mankind throughout Earth.
Although the physical bodies of all humans had
been sentenced to physical death, the spirit of
each human (the breath of God breathed into
Adam) remained immortal. Each human begins
physical life on Earth with a body, soul and

spirit.

The spirit is the eternal breath of God. The soul is the living essence of each individual and is the seat of desires, emotions, and the living characteristics that distinguish each individual from another human. The soul is the means whereby the body interacts with habitat, other humans and the lower hierarchy of living organisms.

The physical body begins to die from birth and is separate from the soul and spirit which are immortal and eternally joined together. Upon physical death of the body it decays back into the elements from which it was created. The soul and spirit being freed from the body await the final judgment of God upon humanity in a place determined in accordance with acceptance or rejection of God's plan for redemption based upon a divine sacrifice.

God determined to allow human souls and spirits to be redeemed through a "sacrificial lamb." Human bodies being cursed and appointed to die are not redeemable. God decided to punish Himself for human trans-gressions by living on Earth as a human in order to offer Himself as every human's sacrificial lamb.

God selected and called out a man named Abraham who honored God and wanted a personal relationship with Him. Abraham was chosen by God to father the lineage through which He would be born of a virgin and live on

Earth as a man subject to the full range of human temptations before offering up His sinless body and blood to redeem human souls and spirits.

Abraham's grandson, Jacob, fathered twelve sons. Jacob's name was changed to Israel. Thus, Israel's twelve sons fathered the "twelve tribes of Israel," who are collectively referred to simply as Hebrews, Jews, Israel, or "God's Chosen People."

God entered into a covenant with Abraham to give him and his descendants a geographical area located in the Middle-East which became known as "the Promised Land." The sign of this covenant with Abraham was circumcision of all his male descendants.

In the Light   Don Alexander

During a severe famine which included the Promised Land, Israel and his lineage moved to Egypt where they flourished and multiplied exceedingly. In process of time, the Egyptians reduced Israel to slavery which lasted four hundred years.

A man named Moses, born an Israelite but raised as an Egyptian prince, was called by God to lead Israel out of slavery. Through Moses, God afflicted the Egyptians mightily until they thrust Israel out whereupon Israel after wandering forty years in the Sinai Desert finally returned to the Promised Land. They conquered the nations inhabiting the land and divided the land among the twelve tribes.

During the desert wanderings, God gave

In the Light   Don Alexander

Moses the Ten Commandments along with a
civil, criminal and ceremonial law. The single
purpose of the law was to make humans aware
of their need for a divine sacrifice. Animal
sacrifices substituted for God's personal offering
up of Himself until such time that the divine
sacrifice would be actually accomplished to
redeem human souls and spirits.

God promised Israel special blessings for
following the law and cursing for disobedience.
Israel drifted into disobedience, idolatry and
debauchery. God sent Israel numerous prophets
who warned Israel of coming chastisement and
dispersion. Israel rejected a theocracy under
God and demanded a king among whom was
King David.

David honored God and was highly favored by Him. God made a covenant with David that his throne would endure forever. David's son, Solomon, succeeded him and was endowed by God with unlimited wisdom and great wealth. In his old age, Solomon's young wives coaxed him into idolatry. His son, Rehoboam, threatened to raise taxes and Israel split into two kingdoms with a monarch over each division. The Northern Kingdom retained the name Israel (ten tribes), and the Southern Kingdom became known as Judah (two tribes).

The succeeding monarchs in both Israel and Judah waged war with each other and with the remnants of the seven nations sharing the Promised Land. All of the monarchs in Israel practiced idolatry but a few in Judah tried to obey God's law. God continued to send prophets to warn Israel and Judah of pending chastisement and dispersion among other nations referred to as "Gentiles."

The prophets also predicted the entire course of human history from the idolatry of Israel and Judah to the end of time. Israel and Judah persecuted the prophets whereupon they were driven from the Promised Land and dispersed among the Gentile nations. Israel was carried away captive into Assyria and lost their distinction, becoming known as "the Ten Lost Tribes." Judah was carried away captive 152 years later into Babylon.

Judah endured seventy years of captivity

under the Babylonians and under the Persians
who overthrew the Babylonians. The Persians
kings allowed all Israelis to return to their
homeland, but only a small remnant returned.

The remnant of Judah was inhabiting the
Promised Land under Roman rule when God,
Himself, appeared on Earth, born of a virgin in
a stable and named Jesus.

Jesus Christ began His public ministry at
the age of thirty and offered Himself up before
God, the Father as humanity's sacrificial lamb
approximately three and a half years later. He
appointed twelve apostles to carry on His
ministry after His death and resurrection.

During His public ministry, He was seen
by and followed by great multitudes of human

observers who heard Him speak and watched
Him open blind eyes, restore withered limbs,
cleanse lepers, heal all types of other sicknesses
and diseases, raise people from the dead, cast
out demons and evil spirits, feed thousands with
a handful of bread and fish, and turn water into
wine.

He said He was God in human form and
would offer up His body and blood to atone for
the sins of all humans and then rise from the
dead. To those who doubted, He encouraged
them to believe on Him because of the mighty
miracles He performed in their sight and
hearing. He cautioned that He was the only
divine sacrifice that would ever be offered up,
and those who reject Him will be forever cut off

from the Godhead and spend eternity in a very unpleasant place.

After being crucified and rising from the dead, Jesus Christ was seen and heard for forty days as He instructed His apostles and disciples regarding their ministry in his footsteps. He was seen by more than five hundred people at the same time. He physically appeared to His apostles on several other occasions before ascending back to the throne of the Godhead.

The enemies of Jesus who were instrumental in His physical death on a Roman cross never disputed His many miracles because there were simply too many eye witnesses. Rather, they contended that He taught against the law of Moses, raised insurrection against

Rome, forbade paying taxes to Rome, was a blasphemer, violated the Sabbath Day, and deceived the people by casting out evil spirits and demons through the power of Satan, himself.

Moreover, the Israeli High Priest, Scribes, Pharisees, Sadducees, and Herodians plotted to kill one of the individuals Jesus raised from the dead because they wished to discredit the man's resurrection.

The Holy Bible further teaches that Hell is inside Earth, that God's throne is in the sides of the North, and that Hell is a prison where the fire is never quenched and those confined there never die. After the final judgment of those who reject Jesus Christ, Hell and its occupants are transferred to an eternal "lake of fire" where Satan, fallen angels, and demons will spend eternity.

The prophecies concerning the end of the dispensation of grace and mercy predict a nuclear war followed by a seven-year reign of Satan (the great tribulation period). During this Satanic dictatorship, deception, torture and murder will be loosed on lip service Christians left behind when true Christians are raptured from Earth. The rapture will occur seven years prior to the Battle of Armageddon which will be interrupted by the second coming of Jesus Christ to Earth.

At His second coming, Jesus destroys all the military forces on Earth and rules over earth for one thousand years. Satan is bound and powerless during this period. The redeemed from Earth (the raptured Christians) will rule with Jesus over the remnants of the nations which survive the great tribulation period, and over the remnant of Israel which escape Satan's genocide campaign against all Jews.

At the conclusion of the millennium reign of Christ on Earth, Satan will be allowed to recruit a vast army from among those humans who enter into the millennium reign in their natural bodies.

Satan leads his army against Jesus and redeemed Christians whereupon he is forever defeated and cast into the lake of fire to spend eternity. The final judgment of those rejecting

In the Light   Don Alexander

Christ follows wherein Hell gives up its
occupants to be formally judged and then be cast
into the lake of fire.

      Thereafter, Earth is cleansed by fire
while the elements melt with a fervent heat.
Redeemed Christians join God's unending
kingdom within a new creation which includes a
new Earth.

CHAPTER   EIGHT

Within the Holy Bible, Genesis describes the reconstruction of Earth; the creation of Earth's life forms including humans; the global flood; the Tower of Babel; the call of Abraham; the lineage of Abraham's descendants through four generations; and resettlement of Abraham's seed in Egypt where they became wealthy, numerous and contented.

Exodus, Leviticus, Numbers and Deuteronomy record the Israeli slavery in Egypt; the freedom journey led by Moses; the wilderness wanderings; the giving of God's law to Moses; the grumbling and complaining by Israel during the forty years in the wilderness; and the death of Moses.

Joshua, Judges, Ruth, First and Second Samuel relate Israeli wars to possess the Promised Land; the original division of the land among the twelve tribes; and the history of Israel between the settlement of the Promised Land and the reign of King David.

First and Second Kings plus First and Second Chronicles record Israel's history from the death of King David to the captivities of both Israel and Judah.

Ezra, Nehemiah, Ester, Ezekiel, and Daniel tell the story of Judah's history during the Babylonian captivity and the return of a remnant of Judah to the Promised Land under Persian rule. Ezekiel and Daniel also spoke and wrote specific prophecies covering future events from Judah's captivity to the millennium reign of Christ. Ezekiel reminds the generations born in captivity that God is punishing Judah for rebelling against Him by idol worship and debauchery.

Job records the personal history of a man named Job who remains faithful to God in the face of unrelenting persecution by Satan.

Psalms is Israel's book of prayers and praises to Jehovah and contains numerous prophecies concerning Jesus Christ.

Proverbs is a collection by King Solomon of wise sayings of his time along with instructions for achieving personal peace and contentment.

Ecclesiastes and the Song of Solomon are the writings of King Solomon concerning his thoughts, beliefs and personal experiences as the wisest and wealthiest individual to reign over all twelve tribes of Israel.

Isaiah, Jeremiah, Lamentations, Hosea, Joel, Amos, Obadiah, Jonah, Micah, Nahum, Zephaniah, Haggai, Zechariah and Malachi are books written by prophets God sent to Israel and Judah.

Their prophecies cover Israeli history from the monarchies in Israel and Judah to the fourth century before the birth of Jesus Christ. The sermons and writings of these prophets repeatedly warn Israel and Judah to repent of idolatry and wicked practices to avoid chastisement and dispersion among Gentile nations. These same prophets also predicted hundreds of future events spanning human history on Earth from 780 B.C. to the end of time.

The Holy Bible is silent from the prophecies of Malachi to the birth of Christ (400 years). In the New Testament portion of the Holy Bible; Matthew, Mark, Luke and John relate the birth, ministry, sacrificial death and resurrection of Jesus Christ.

The book of Acts covers the ministries and persecutions of the apostles of Christ during their teaching of Christianity throughout the Roman Empire.

Romans, First and Second Corinthians,
Galatians, Ephesians, Philippians, Colossians,
First and Second Thessalonians, First and
Second Timothy, Titus, and Philemon, trace the
ministry of Paul, the Apostle to the Gentiles
through Paul's letters to the churches which he
founded.

The emphasis is on redemption back to
God through the divine sacrifice of Jesus Christ
rather than through the Law given to Moses.
Paul explains that the whole purpose of the Law
is to point humans to their need for redemption
through a God-given sacrificial lamb.

The book of Hebrews summarizes the
distinctions between God's Law and God's
Grace. James, First and Second Peter, First,

Second and Third John, and Jude are letters written by the apostles Peter and John and by Jude, the brother of James. The letters are directed to individuals, believers and non-believers setting forth basics Christian principles and offering encouragement to those seeking to practice the teachings of Jesus Christ.

Revelation, the last book in the Holy Bible, was written by the apostle John during a period of imprisonment on the Isle of Patmos in 96 A.D. Revelation looks back at what has been, what is in the immediate future and then what is going to happen from the apostolic age to the end of time including the cleansing of Earth and God's new creation.

John's vision on Patmos along with

selected passages from the King James Version of the Holy Bible (which is "public domain") describes the great tribulation period and Satan's final efforts to deceive humanity:

When the rapture of Christians occurs, the people who remain on Earth will be aware that certain Christians are missing, but will be unaware of the true reason therefor. Excuses will be readily offered to explain away the missing individuals who will represent a small percentage of Earth's human population.

Life on Earth will continue and the raptured Christians will soon be old news as those left behind are captivated by emergence of a new celebrity (the Antichrist). This magnetic personality will provide believable solutions to previously insolvable problems and will orchestrate global peace including a treaty with Abraham's seed (Israel).

Pursuant to being accepted as dictator over all humanity, the Antichrist will use the absolute power given him to wage a genocide campaign against all Jews and to hunt down the "divine vomit" (left behind, lukewarm Christians) who will recognize the Antichrist as Satan in the flesh (the beast).

The treaty with Israel will be violated and the beast will proclaim himself to be Almighty God. Peace will revert to global war ushering in pestilence, famine, and decimation of Earth's population.

These events are revealed to John during

the portion of his Patmos vision pertaining to "the things which shall be hereafter." A scroll, written on both sides and sealed with seven seals, is being unsealed by "the Lamb of God" (Jesus). The angels are in attendance along with four divinely ordained living creatures (the four beasts), and twenty-four "elders" from among the raptured Christians and the redeemed remnant of Israel.

"............the Lamb opened one of the seals, and I heard, as it were the noise of thunder, one of the four beasts saying, Come and see. And I saw, and behold a white horse: and he that sat on him had a bow; and a crown was given unto him: and he went forth conquering, and to conquer. And when He had

opened the second seal, I heard the second beast
say, Come and see. And there went out another
horse that was red: and power was given to him
that sat thereon to take peace from the earth, and
that they should kill one another: and there was
given unto him a great sword. And when He
had opened the third seal, I heard the third beast
say, Come and see. And I beheld, and lo a black
horse; and he that sat on him had a pair of
balances in his hand. And I heard a voice in the
midst of the four beasts say, A measure of
wheat for a penny, and three measures of barley
for penny; and see thou hurt not the oil and the
wine. And when He had opened the fourth seal,
I heard the voice of the fourth beast say, Come
and see. And I looked, and behold a pale horse:

and his name that sat on him was Death, and
Hell followed with him. And power was given
unto them over the fourth part of the earth, to
kill with sword, and with hunger, and with
death, and with the beasts of the earth."
(Revelation, chapters 4, a portion of verses 1-11;
chapter 6, verses 1-8, Patmos vision, 96 A.D.)

The nuclear war described in Zechariah,
chapter 14, verse 12; and in Ezekiel, chapter 38,
verses 1-23; chapter 39, verses 1-16 will set
Israel apart from other nations to such an extent
that the Antichrist will have to negotiate a peace
treaty with the Jews in order to bring about the
false peace which vaunts him into power as
global dictator.

This false peace will last forty-two

months (the first half of the reign of Satan on
Earth). The peace will be broken when the beast
(Antichrist) sets out to exterminate the seed of
Abraham:

"And woe unto them that are with child,
and to them that give suck in those days! But
pray ye that your flight be not in winter, neither
on the sabbath day: for then shall be great
tribulation, such as was not since the beginning
of the world to this time, no, nor ever shall be.
And except those days should be shortened,
there should no flesh be saved: but for the elect's
sake those days shall be shortened." (Matthew,
chapter 24, verses 19-22; Jesus Christ
prophesying to his disciples)

"And there followed him a great

company of people, and of women, which
bewailed and lamented him. But Jesus turning
unto them said, Daughters of Jerusalem, weep
not for Me, but weep for yourselves, and for
your children. For, behold, the days are coming,
in the which they shall say, Blessed are the
barren, and the wombs that never bare, and the
paps which never gave suck. Then shall they
begin to say to the mountains, Fall on us; and to
the hills, Cover us. For if they do these things in
a green tree, what shall be done in the dry?"
(Luke, chapter 23, verses 27-31; spoken by
Jesus on the way to his sacrificial death)

In 538 B.C., the angel, Gabriel,
explained the treachery of Satan in the flesh
(Antichrist) to Daniel the prophet. The time

references were given in seventy "weeks of years" (one week equals seven years; or a total period of time equal to seventy times seven years):

"Seventy weeks are determined upon thy people and upon thy holy city, to finish the transgression, and to make an end of sins, and to make reconciliation for iniquity, and to bring in everlasting righteousness, and to seal up the vision and prophecy, and to anoint the most Holy. Know therefore and understand, that from the going forth of the commandment to restore and to build Jerusalem unto Messiah the Prince shall be seven weeks, and threescore and two weeks: the street shall be built again, and the wall, even in troublous times. After threescore

and two weeks shall Messiah be cut off, but not
for Himself: and the people of the prince that
shall come shall destroy the city and the
sanctuary; and the end thereof shall be with a
flood, and unto the end of the war desolations
are determined. And he shall confirm the
covenant with many for one week: and in the
midst of the week he shall cause the sacrifice
and the oblation to cease, and for the
overspreading of abominations he shall make it
desolate......." (Daniel, chapter 9, verses 24-27;
spoken by Gabriel, 538 B.C.)

Four hundred and eighty-three years
(sixty-nine weeks of years) elapsed between the
decree allowing the remnant of Judah and
Benjamin to return to Jerusalem, and the death

of Christ, exactly as predicted by Gabriel.

One week of years (seven years) still remain to be fulfilled as referenced by Gabriel to Daniel in 538 B.C. This period of seven years will be fulfilled during the reign of the Anti-Christ, and the clock will begin ticking again following the rapture of faithful Christians from Earth. The first three and one half years of Antichrist's reign will be peaceful, and the final three and one half years will be filled with terror, torture, murder, war, and genocide:

"And one said to the man clothed in linen, which was upon the waters of the river, How long shall it be to the end of these wonders? And I heard the man clothed in linen, which was upon the waters of the river, when he

held up his right hand and his left hand unto
heaven, and sware by Him who liveth forever
that it shall be for a time, times, and a half; and
when he shall have accomplished to scatter the
power of the holy people, all these things shall
be finished. And I heard, but I understood not:
then said I, O my Lord, what shall be the end of
these things? And he said, Go thy way, Daniel:
for the words are closed up and sealed till the
time of the end. Many shall be purified, and
made white, and tried; but the wicked shall do
wickedly; and none of the wicked shall
understand; but the wise shall understand. And
from the time that the daily sacrifice shall be
taken away, and the abomination that maketh
desolate set up, there shall be a thousand two

hundred and ninety days."   (Daniel, chapter 12, verses 6-11)

The abomination that maketh desolate is the violation of the Jewish temple by Antichrist wherein Satan, in the flesh, occupies the temple and proclaims himself to be Almighty God.

This occurs at the midpoint of his seven year reign, and coincides with his efforts to kill every Jew, and to torture to death all who refuse to take his mark and to worship him as god. He will wage war against Israel, and roving death squads will hunt down the "divine vomit" (those lukewarm Christians left behind at the rapture, but knowing enough to reject Satan's mark and to refuse to worship him):

"And he had power to give life unto the

image of the beast, that the image of the beast
should both speak, and cause that as many as
would not worship the image of the beast should
be killed. And he causeth all, both small and
great, rich and poor, free and bond, to receive a
mark in their right hand, or in their foreheads:
And that no man might buy or sell, save he had
the mark, or the name of the beast, or the
number of his name. Here is wisdom. Let him
that hath under standing count the number of the
beast: for it is the number of a man; and his
number is six hundred threescore and six."
(Revelation, chapter 13, verses 15-18; Patmos
vision, 96 A.D.)

To understand the rotating references to
Satan, Antichrist, and the false prophet, it is

helpful to remember that Satan is the spiritual power sustaining his incarnation in the body of the individual identified as the Antichrist, whereas the false prophet is the Satanic substitute for the Holy Spirit. The false prophet, acting as the world's spiritual figurehead, calls upon all humanity to worship the beast (Antichrist).

During the last forty-two months (3 1/2 years, 1,260 days) of Antichrist's reign, two witnesses oppose him and, like Moses opposing Pharaoh, call upon God to plague the kingdom of Antichrist and his followers. This period is also referred to as the "great tribulation."

The population of Earth must endure both the wrath of God and the terror perpetuated by Antichrist. The devastating plagues pursuant to the ministry of the two witnesses will be similar in nature to the plagues suffered by Egypt during the time of Moses, but of much greater intensity.

The mayhem, torture and murder wrought by Antichrist will be of such magnitude that the majority of Jews will be killed, and over half of non-Jews will be slaughtered.

One hundred and forty-four thousand Jewish male virgins will be protected by God and serve as evangelists during the great tribulation. Millions will be martyred rather than worship Satan:

"After this I beheld, and lo, a great multitude, which no man could number of all nations, and kindreds, and people, and tongues, stood before the throne, and before the Lamb, clothed with white robes and palms in their hands; And cried with a loud voice, saying, Salvation to our God which sitteth upon the throne, and unto the Lamb........And one of the elders answered, saying unto me, What are these which are arrayed in white robes? and whence came they? And I said unto him, Sir, thou knowest. And he said unto me, These are they which came out of great tribulation, and have washed their robes, and made them white in the blood of the Lamb. Therefore are they before the throne of God, and serve Him day and night in His temple: and He that sitteth on the throne shall dwell among them." (Revelation, chapter

7, verses 9-10 and 13-15)

Toward the end of the second half of Antichrist's reign, an alliance of nations will rally themselves against Antichrist and his armies. The battle will commence in the Middle East at a place called Armageddon. Antichrist will be sitting in the Jewish temple as God Almighty.

Armies opposing Antichrist will be attempting to dethrone him, and the armies supporting Antichrist will be defending his claim to divinity. The surviving Jews and the city of Jerusalem will be caught in the middle of the conflict. This final battle between humans orchestrated by Antichrist will be ended by the second coming of Jesus Christ:

In the Light   Don Alexander

"And I saw heaven opened, and behold a
white horse; and He that sat upon him was
called Faithful and True, and in righteousness
He doth judge and make war. His eyes were as
a flame of fire, and on His head were many
crowns; and He had a name written, that no man
knew, but He Himself. And He was clothed
with a vesture dipped in blood: and His name is
called The Word of God. And the armies which
were in heaven followed Him upon white
horses, clothed in fine linen, white and clean.
And out of His mouth goeth a sharp sword, that
with it He should smite the nations: and He
shall rule them with a rod of iron: and He
treadeth the wine press of the fierceness and
wrath of Almighty God.......And the beast was

taken, and with him the false prophet that wrought miracles before him, with which he deceived them that had received the mark of the beast, and them that worshiped his image. These both were cast alive into a lake of fire burning with brimstone. And the remnant were slain with the sword of Him that sat upon the horse, which sword proceeded out of His mouth (His spoken word): and all the fowls were filled with their flesh." (Revelation, chapter 19, verses 11-15; 20-21)

The Battle of Armageddon is followed by the reign of Christ on Earth which will encompass a period of one thousand years. The beast and the false prophet are imprisoned within the lake of fire, but Satan, himself, is

146

bound and powerless during this period known
as the "millennium reign of Christ."

Earth,   as we know it, will not be
destroyed until the millennium reign is over, and
the promises God made to King David, to
Abraham, and to faithful Christians have been
fulfilled by Jesus Christ sitting upon the throne
of David. Thereafter, Satan will be loosed and
allowed to wage his final battle against God:

"And I saw an angel come down from
heaven, having the key of the bottomless pit and
a great chain in his hand. And he laid hold on
the dragon, that old serpent, which is the Devil,
and Satan, and bound him a thousand years, and
cast him into the bottomless pit, and shut him
up, and set a seal over him, that he deceive the

nations no more, till the thousand years should be fulfilled: and after that he must be loosed a little season. And I saw thrones, and they sat upon them, and judgment was given unto them: and I saw the souls of them that were beheaded for the witness of Jesus, and for the word of God, and which had not worshiped the beast, neither his image, neither had received his mark upon their foreheads, or in their hands; and they lived and reigned with Christ a thousand years. But the rest of the dead lived not again until the thousand years were finished. This is the first resurrection. Blessed and holy is he that hath part in the first resurrection: on such the second death hath no power, but they shall be priests of God and of Christ, and shall reign with Him a thousand years. And when the thousand years are expired, Satan shall be loosed out of his prison, and shall go out to deceive the nations which are in the four quarters of the earth, Gog and Magog, to gather them together to battle: the number of whom is as the sand of the sea. And they went up on the breadth of the earth, and encompassed the camp of the saints about, and the beloved city: and fire came down from God out of heaven and devoured them. And the Devil who deceived them was cast into the lake of fire and brimstone, where the beast and the false prophet are, and shall be tormented day and night for ever and ever." (Revelation, chapter 20, verses 1-10; Patmos vision, 96 A.D.)

The consignment of Satan to the lake of

fire is followed by the "great white throne" judgment. The only individuals who appear at this judgment are those who steadfastly refused to accept God's forgiveness through the divine sacrifice of Jesus Christ:

"And I saw a great white throne, and Him that sat on it, from whose face the earth and the heaven fled away; and there was found no place for them. And I saw the dead, small and great stand before God; and the books were opened: and another book was opened, which is the book of life: and the dead were judged out of those things which were written in the books, according to their works. And the sea gave up the dead which were in it; and death and hell delivered up the dead which were in them: and they were judged every man according to their works. And death and hell were cast into the lake of fire. This is the second death. And whosoever was not found written in the book of life was cast into the lake of fire." (Revelation, chapter 20, verses 11-15)

This then is the death of all corrupted species upon the planet, Earth. The redeemed among all humanity will be in heaven with their redeemer, Jesus Christ. Those refusing redemption will be forever in the lake of fire with their spiritual father, Satan. Then will be brought to pass that which is written:

"Looking for and hasting unto the coming of the day of God, wherein the heavens being on fire shall be dissolved, and the

elements shall melt with fervent heat ....
Nevertheless we, according to his promise, look
for new heavens and a new earth, wherein
dwelleth righteousness." (II Peter, chapter 3,
verses 12-13)

      "And I saw a new heaven and a new
earth: for the first heaven and the first earth
were passed away; and there was no more sea.
And I, John, saw the holy city, new Jerusalem,
coming down from God out of heaven, prepared
as a bride adorned for her husband."
(Revelation, chapter 21, verses 1-2)

CHAPTER   NINE

When summarizing world religions (other than Christianity), there are twelve that merit separate description. All other religions or belief systems which reject Jesus Christ as the sacrificial lamb of God are little more than simplistic variations of Islam, Hinduism, Buddhism, Bahai, Judaism, Zoroastrianism, Taoism, Unitarian Universalists, Jainism, Evolution, and Scientology.

Islam, with 1.5 billion followers, is the second largest religion on Earth (after Christianity with 2.1 billion followers). Islam was founded more than 500 years after the resurrection of Christ by an Arab named Muhammad who regarded himself as Allah's (Arabic for God's) guardian of the true faith of Abraham.

Muhammad proclaimed that Jews and Christians distorted the revelations God gave to Abraham, Moses, Jesus and other prophets by text altering and misinterpretation. The sacred writings of Islam are referred to as the Qur'an (God's personal revelations to Muhammad) and the Sunnah (words and deed of Muhammad, God's final prophet to Earth).

Followers of Islam are known as Muslims who deny that God had a son. They believe that Jesus was a mere prophet, that he escaped into Paradise and did not sacrifice himself for the sins of all humans. Everyone must save themselves from hell by accumulating more good works than evil works, and true believers may have to spend some time in hell to atone for insufficient good works compared to evil works.

Upon birth, an individual's record is opened in Paradise and the individual becomes personally chargeable upon reaching the age of accountability (puberty). It is permissible to lie, steal, kill, rob, rape and pillage in the service of Allah (converting infidels from the error of their ways).

Paradise is a place of feasting, drinking, and sexual gratification surrounded by a swarm of virgins. Muslims who die or commit suicide in service to Allah are ushered directly into Paradise. Muslims condemn homosexuality, adultery, eating pork and gambling.

Islam is the by far the most intolerant and most violent "save yourself" belief system in existence.

Zoroastrianism is a religion with approximately 200,000 followers which preaches "save yourself" through good thoughts, good words, and good deeds. Zoroastrianism is also referred to as Mazdaism and first appears in recorded history during the seventh century BC.

Zoroaster is the main prophet for the belief system and invented the term "Ahura Mazda" to name the one universal and transcendental god. Mazdaism embraces the concepts of good and evil. There exists an immortal adversary of Ahura Mazda dedicated to evil (Druj).

Humanity is drawn into the conflict wherein Ahura Mazda is ultimately victorious and time ends with the renovation of the universe. Thereafter, all creation is reunited in Ahura Mazda. The collection of sacred texts are called "Avesta."

Good thoughts, words and deeds are required to ensure happiness and ward off chaos. Today, followers of Zoroastrianism are located primarily in Iran, India and Pakistan.

Unitarian Universalists have around 800,000 followers who promote world unity and the inherent goodness of humans. All religions are embraced and are considered to be of equal merit. The ultimate achievement is religious unity and a single world government. Peace through good thoughts, good deeds and unbiased tolerance is the main tenet.

Otherwise, all followers are encouraged to worship as they choose. Consequently, Unitarian Universalists are everything to everybody and humans will eventually live together happily as quickly as unity becomes reality. Abortion, homosexuality, and same sex marriage are smiled upon.

A fiction writer, L. Ron Hubbard, gave birth in 1960 to what has become known as Scientology, with roughly a half million adherents, and described as the study and handling of the human spirit in relationship to itself, others, and all life. The sacred texts are various books written by Hubbard.

Man, a gender neutral term which refers to all humans, is basically good, but life experiences lead him into evil. Man errs by trying to solve his problems from his own point of view rather than achieving greater spiritual awareness through learning, auditing and training. Man is a spiritual being whose existence spans more than one lifetime.

Man is endowed with abilities well beyond those he normally considers he possesses. What is true for man is what he has observed to be true. During each reincarnation, man applies the knowledge and increased spiritual awareness he acquired during the previous life.

Man can improve his quality of life to the degree he continues to preserve his spiritual integrity and remains honest and decent thus achieving certainty of spiritual existence and a relationship with whatever supreme being he believes exists outside of himself.

Scientology organizations provide ongoing auditing and counseling. Because man alone controls his earthly and eternal existence, Scientology is a somewhat unusual "save yourself" belief system with Hubbard's home spun psychiatry and hypnosis imbedded within the pseudo-scientific orientation.

In the Light   Don Alexander

The seven million Bahai followers
believe in one god who created everything; but
the Bahai god is transcendent and unknowable
who has and will continue to send great prophets
to humanity through which the unknown deity
has revealed a series of messages. Bahai
prophets thus far have been Adam, Abraham,
Moses, Krishna, Zoroaster, Buddha, Jesus
Christ, Mohammed, The Bab, and Bahaullah.

Another prophet is not expected for
many centuries into the future. Bahai teaches the
essential unity of the great world religions as
arising from the same spiritual source but
splintered by conditions at the time of founding
and by accretions following the death of the
founder.

In the Light   Don Alexander

The Bahai faithful believe that all
individuals possess an immortal soul not subject
to decomposition. At death, the soul is freed to
travel throughout the spiritual universe which is
a timeless and place-less extension of the known
universe. The sacred texts are a collection of the
writings of Abdul--Baha, The Bab, and
Bahaullah plus miscellaneous Bahai scriptures
which were first circulated during the nineteenth
century A.D.

Bahai further teaches that the happiness
of mankind as well as world peace and security
are unattainable until global unity is firmly
established. Bahai promotes gender and race
equality, world government, freedom of
expression and assembly, world peace, religious

tolerance, and religious cooperation.

Bahai rejects homosexuality while calling for equal dignity and respect for all peoples, the elimination of poverty and excessive wealth, universal education and economic justice. Mankind must control the present and future through unity and global cooperation.

The sacred texts of nine million adherents to Hinduism are collectively referred to as "the Vedas" and the written forms date between 600 to 300 B.C. Hinduism views the entire universe as having one divine entity who is at one with the universe but transcends it as well. Brahma is the creator who is always creating new realities. Dharma is the eternal order, religion, law and duty. Vishnu preserves the creations of Brahma and travels between heaven and earth in one of ten incarnations.

Shiva is the destroyer of eternal order but can be compassionate and erotic. Hinduism is splintered into various groupings which worship local gods and goddesses.

The two major divisions of Hinduism are Vaishnavaism (Vishnu is the ultimate deity) and Shivaism (Shiva is the ultimate deity). The main tenet of Hinduism is the transmigration of the soul -- a continuous cycle of birth, life, death, and rebirth through many lifetimes (referred to as "samsara").

Hindu priests serve at rituals and worship ceremonies but are considered unnecessary in rural areas where priestly duties are carried out by local non-Brahmins.

The four aims of Hinduism or "the doctrine of the fourfold end of life" are Dharma (religious righteousness), Artha (economic success and wealth), Kama (gratification of the senses such as sex, pleasure, and mental enjoyment), and Moska (liberation from samsara).

The three goals of the "pravritti" (those who are in the world) are Dharma, Artha, and Kama. The main goal for the "nivritta" (those who renounce the world) is Moska. Liberation from samsara, thus becoming one with the universe, is the supreme goal of mankind.

Kama also refers to the accumulation of

an individual's good and bad deeds. An overload of bad Kama might result in rebirth as an animal or insect. The unequal distribution of wealth, prestige, and suffering is believed to be the result of one's previous acts during the current life and previous lives.

Meditation is practiced with Yoga being the most observed. Other Hindu activities include rituals, daily prayers, and ceremonial dinners for various deities.

Hinduism is a "save yourself" belief system where good thoughts, good intentions, and good deeds will ultimately be rewarded with cessation of reincarnation.

Buddhism, with 376 million followers, was founded during the second half of the sixth century B.C. by Siddhartha Gautama, son of King Gautama who ruled over a small district in the Himalayas between India and Nepal. As a young man, Siddhartha wandered outside the palace and observed a leper, a corpse, and an ascetic whereupon he concluded that happiness is an illusion.

After he fathered a son to ensure the royal bloodline, Siddhartha began a pilgrimage of inquiry and asceticism wherein he was influenced by two Brahmin hermits and later by five monks. After years of seeking communion with the supreme cosmic spirit, Siddhartha claimed to have discovered the four noble truths (Pativedhanana) and pronounced himself the Buddha.

In the Light   Don Alexander

He labored some forty years spreading
the Buddha doctrines and died a questionable
death at age 80 (it is reported that he was
poisoned by a blacksmith). The teachings of
Buddha are referred to as Dharma. Following
Siddhartha's death, his followers convened to
create tenets they could all accept within the
caste system which requires a series of rebirths
to move up through the system.

The Buddha rejected the concepts of a
supreme being and eternal souls. Whatever gods
inhabit the cosmos are impermanent and are
reincarnated like humans.

The cessation of rebirths is named
"nirvana" wherein the individual being becomes
one with the Universal Soul. Nirvana is the

ultimate achievement.

Karma (tally of good and bad deeds) determines the kind of rebirth and quality of life after rebirth. The path to nirvana is to follow the four noble truths -- the universality of suffering; the origin of suffering; overcoming of suffering; and the suppression of suffering. Lustful desires cause suffering which is experienced during rebirth, aging, death and rebirth. Suffering can be overcome by suppression of the desires causing one to suffer.

The way leading to suppression of suffering is a noble path with eight branches -- right aspirations, right speech, right conduct, right livelihood, right effort, right concentration, right views of understanding, and right mindfulness. The eight branches are different dimensions of a total way of life.

Several lives are required to achieve nirvana. The journey is long and difficult with inner peace and harmony as one approaches nirvana; then nothingness. The sacred texts of Buddhism were compiled around 80 B.C. and are referred to as the Pali Canon (also called the Tripitaka). In summary, Buddhism rejects the concept of a supreme being and teaches that human works are disciplined by cycles of reincarnation.

Jainism, a heretical movement within Hinduism, was founded by a man named Mahavira. The sacred texts of 4.2 million adherents to Jainism are the twelve "angas" plus lesser writings which appeared in written form around 1600 A.D. At age 30, it is believed by followers of Jainism that Mahavira decided to live a life of self-denial and wandered naked through India for twelve years before achieving "enlightenment."

In his thirteenth year of naked wandering, in a squatting position exposed to the sun with his knees high and his head low, in deep meditation, Mahavira reached nirvana whereupon he stopped living by himself and attracted disciples. He preached his revelations until his death at which time he allegedly boasted of 14,000 monks within his brotherhood.

Although Mahavira was steadfastly opposed to the concept of God or gods, his followers elevated him to deity claiming that he descended from heaven without sin and having all knowledge. Jainism preaches self denial as the path to nirvana.

The Five Great Vows renounce killing any living thing; lying, greed, sexual pleasure, and worldly attachments. Monks were taught to avoid women entirely because Mahavira believed they were the cause of all types of evil.

Taoism traces its roots to Lao-Tzu around 600 B.C. The sacred text is Tao-te-Ching (also known as Daodejing). Taoism was originally a hodgepodge of psychology and philosophy that became a religion in 440 A.D. and benefited from state support until the fall of the Ching Dynasty in 1911 A.D. Today, Taoism has roughly twenty million followers centered mainly in Taiwan off the mainland of China.

Taoism rejects the concept of a personalized deity. Tao is the life force which flows through the universe and all life. The ultimate achievement is to harmonize with Tao. Taoists believe in letting nature take its course unimpeded by mankind.

Time is cyclical and not linear. Kindness is always reciprocated. Left to themselves, people will be compassionate without expecting a reward.

Tao regulates and balances natural processes and embodies the harmony of opposites -- no love without hate; no light without dark; no male without female, etc.

In the Light   Don Alexander

There is no God to hear prayers or to act on
them. Taoists seek answers to life's questions
through inner meditation and outer observation.

Evolution is the religion of atheists
who deny that any supreme being exits; and that
all life forms evolved from a single living cell
which created itself through a chain of unrelated
and purely random events. From this original
living cell evolved every life form that exists
today or has ever existed in the past eons of
time.

The elements within the universe
emerged from nothingness by random chance
for no reason and without purpose. There is no
life of any kind following physical death.

Humanity is simply an accidental life

form produced by natural selection and survival of the fittest. The number of atheists is estimated world-wide at approximately 1.1 billion. The religion of "Evolution" or "Spontaneous Generation" which denies any form of design or creation dates back to around 800 B.C.

Judaism is a belief system traced back to a man called Abraham who lived in Ur of the Chaldees within the fertile crescent around 1900 B.C. Although his family worshiped idols, Abraham believed there is an unseen supreme being who created the universe and all life forms.

According to Judaism, Abraham meditated upon and prayed to the invisible God until God called him out from among the idol worshipers and instructed him to travel to a land which God would give to him and to his seed. The "promised land" would be shown to him as he traveled.

Abraham obeyed God and went out not knowing where he was going. God led him to the land of Canaan where Abraham lived as a shepherd and fathered Ishmael and Isaac. Ishmael was the firstborn, but his mother was a bond-servant. Isaac's mother was Sarah, Abraham's wife.

Ishmael and his seed fathered the Arabs; and Isaac's descendants became known as Jews. Isaac's wife birthed two sons named Jacob and Esau. Jacob's name was changed to Israel and his twelve sons begat twelve tribes who

collectively are referred to as "the children of
Israel." Esau's descendants are called Edomites.
Esau, being the firstborn, sold his birthright to
Jacob for a bowl of stew.

The sacred texts of Judaism are the thirty-
nine narratives which make up the total writings
within the Old Testament of the Holy Bible.
Followers of Judaism believe in one true God,
sin and righteousness, resurrection from the
dead, heaven and hell, Satan and the angels,
eternal life, and eternal punishment.

They further believe in sacrificial
offerings (animal sacrifices) to obtain forgiveness for breaking God's law. Moses, the
greatest of God's prophets, received God's law
while seeking God on Mount Sinai around 1491

B.C. The Law of Moses covers criminal, civil and religious law. Religious law involves a priesthood, rituals, sacrifices, holy days, annual feasts, an annual Day of Atonement, a Sabbath day (every 7th day), the Sabbath year (every 7th year), and the year of Jubilee (every 50th year wherein all bond servants are freed and all real estate reverts to the original tribal family).

The ten main points of the Law of Moses are referred to as God's Ten Commandments -- worship God only; do not set up nor worship any graven image; do not speak of God in an irreverent manner; remember and keep the Sabbath day; do not commit adultery; do not steal; do not murder; do not bear false witness; do not covet; and honor both father and mother.

The thirty-nine Old Testament narratives were written and compiled between 1491 and 397 B.C. by a total of thirty-one different authors. The first five narratives are ascribed to Moses and are referred to as the "Pentateuch" (Genesis, Exodus, Leviticus, Numbers and Deuteronomy).

Followers of Judaism who rejected the teachings of Jesus Christ are still waiting for their prophesied Messiah who the Hebrew prophets said would restore Israel to the glory the nation enjoyed under King David and would make Jerusalem the center of world government.

Jews (children of Israel) have been the most hated and persecuted of all nationalities (anti-Semitism) since 825 B.C. and continuing to the present time. Following approximately 2,000 years of dispersion among Gentile nations, the Jews began returning to their homeland pursuant to a United Nations mandate issued in 1948 A.D. During the same year, the tiny nation of Israel declared its independence.

The nation of Israel has remained independent and has become one of the worlds most lethal military powers.

With reference to all world religions other than Judaism and Christianity, there is not a single fact to support the various belief systems. Reincarnation was plucked from the imagination of humans. There has never been a documented case of reincarnation.

The basic goodness of mankind has been

repeatedly proven to be an illusion along with global unity. Good thoughts, good words and good deeds may be exhibited from time to time due to will worship and voluntary self-denial, but do not represent routine human behavior.

Mankind generally exhibits greed, selfishness, envy, cruelty, and hatred for the socially outcast. Past and present wars are very obvious examples of true human nature.

The Law of Cause and Effect eliminates reincarnation as being factual. The basic goodness of humanity has never been demonstrated within relationships between the races, nor between general populations, nor between nations; and very seldom between individuals. Bastardized versions of Christianity, such as

Islam, seek to justify greed, lust. and violence.

Christianity is the only world religion that is supported by facts rather than pure imagination. It has never been reported by hundreds of eye witnesses that they actually saw any of the deities worshiped by non-Christians restore withered limbs, heal all manner of sicknesses and diseases, open the eyes of people born blind, cleanse lepers, raise the dead, or demonstrate control over storms and walk on the water.

Neither has there ever lived on Earth any individual other than Jesus Christ who exhibited total knowledge of all things and who put His most vicious enemies to complete silence.

Secular history testifies to the birth,

ministry, mighty miracles, death by crucifiction, resurrection, and forty-days post-resurrection ministry of Jesus Christ. Those who refuse to believe in Him are like the Chief Priests, Elders and Scribes who admitted Jesus raised the dead and wanted to kill the resurrected individuals to destroy the evidence of such a wondrous work by someone who threatened their traditional belief system.

The Holy Bible has been attacked more viciously and more continuously than any other book ever written. For more than two thousand years the Holy Bible has proven unassailable in terms of prophesied events, historical references, scientific challenges, explanations for human motives and behavior, and the revelation of God's existence and character reflected in the person of Jesus Christ, God's Incarnate Word.

CHAPTER   TEN

The Holy Bible teaches that redemption of human souls and spirits back to God is based upon divine grace (God's unmerited favor) rather than self-sacrificing, self-denial or accumulated "good works." The one and only path to redemption is through the body and blood of Jesus Christ. Personal works have zero to do with redemption (personal salvation) back to God.

Herein lies the contrast between salvation through grace alone, versus salvation through faith plus works. Faithful stewardship earns a reward….not salvation. For it is proclaimed by Jesus through the apostle, Paul, that "with the heart man believeth unto righteousness; and with the mouth confession is made unto salvation." (Romans 10:10)

An individual is "born again" (a spiritual rebirth) such that God becomes his/her spiritual father (rather than Satan, the master of fallen angels) through the exercise of child-like faith in the sacrificial death of Jesus Christ as the substitute for his/her sins…..period! This is the "gift" of God which He extends to every human on Earth as explained repeatedly throughout the entire New Testament Scriptures.

Those who are "born again" become new living creatures and their names are recorded in heaven within the "Book of Life." As "new creatures in Christ," the "born again" are still confined (until death or the second coming of Christ) within physical bodies that are very much subject to being overcome with the lust and cares of this world.

Christian stewardship is measured by how successfully a born again believer struggles against the lust of the flesh, the lust of the eyes, and the pride of life. We <u>earn</u> a heavenly reward for stewardship, and a reward is not a free gift. Salvation is free, whereas rewards are earned totally apart from salvation in accordance with our works during our life on Earth.

Born again Christians do not have God's free gift of salvation ripped away from them because of their transgressions of the law of God given to Moses. Jesus Christ came into this world and sacrificed himself to free humanity from "the law" given to Moses and the consequences of disobedience under that law.

Born again Christians strive to think, say, and do that which the Bible teaches is pleasing to God, their heavenly Father.   Sins are transgressions of God's law, and in the absence of such law,   there can be no sin. Those who believe in and accept Jesus Christ cannot sin because they are no longer subject to God's law as given to Moses. We love and reverence God because He loves us and gave Himself for us through His incarnation in the person of Jesus Christ.

All who refuse to believe in and accept Jesus Christ are still condemned under God's law given to Moses and will spend eternity with their chosen spiritual father, Satan.

God's law given through Moses to humanity was not in effect until 1491 B.C.; which means humans were not guilty of any sins prior to 1491 B.C. -- because sin is by Biblical definition the transgression of God's law. There was "God consciousness" through the breath God breathed into Adam.

Both Adam and Eve lost, through willing disobedience, their spiritual perfection and physical immortality and thus could pass on to their children only a rebellious spirit and mortal bodies. Humanity desperately needed a redeemer (savior) because under God's law no human has or ever will be justified, and we are all found guilty under such laws.

Divine law was given to make humans aware of their need for redemption; and to point us to Jesus Christ who would come into the world and sacrifice Himself as "the Lamb of God" and thus fulfill the exact purpose of the law being given.

Having fulfilled the express purpose of pointing lost humanity to Jesus Christ, the law became null and void as applied to any person believing in, and accepting Jesus Christ as his/her personal savior. Such belief and acceptance triggers a "new birth" and eternal life in God's presence.

Because God is our impartial judge, all humans who lived and died prior to the first coming of Jesus Christ, and all those who never had the opportunity to accept Him as their personal Lord and Savior (for whatever reason…..aborted, mentally incompetent, did not live to the age of accountability, never heard the Gospel of Jesus Christ, etc.) will have a chance to choose between Jesus Christ and Satan. When and how such a choice will be made is not revealed within the Holy Bible and is thus one of many secrets known only to God.

BIBLIOGRAPHY

All Biblical references and quotations, whether a direct quote or paraphrasing, is from the 1611 A.D. Authorized King James Version of the Holy Bible which is public domain.

Carbon-14 Dating, Radiometric Dating and Tree Ring Dating

1. Plastino, W.; Kaih^ola, L.; Bartolomei, P.; Bella, F. (2001). "Cosmic Background Reduction In The Radiocarbon Measurement By Scintillation Spectrometry At The Underground Laboratory Of Gran Sasso". *Radiocarbon* **43** (2A): 157–161. https://digitalcommons.library.arizona.edu/objectviewer?o=http%3A%2F%2Fradiocarbon.library.arizona.edu

%2Fvolume43%2Fnumber2A
%2Fazu_radiocarbon_v43_n2a_157_161
_v.pdf.
2. ^ Arnold, J. R.; Libby, W. F. (1949).
"Age Determinations by Radiocarbon
Content: Checks with Samples of Known
Age". _Science_ **110** (2869): 678–680.
doi:10.1126/science.110.2869.678.
PMID 15407879.
http://hbar.phys.msu.ru/gorm/fomenko/li
bby.htm.
3. ^ Willard Frank Libby
4. ^ _a_ _b_ _c_ Münnich KO, Östlund HG, de
Vries H (1958). "Carbon-14 Activity
during the past 5,000 Years". _Nature_ **182**
(4647):   1432–3.
doi:10.1038/1821432a0.
5. ^ _a_ _b_ Ramsey, C. Bronk (2008).
"Radiocarbon dating: revolutions in
understanding". _Archaeometry_ **50** (2):
249-275.   doi:10.1111.2Fj.1475-
4754.2008.00394.x. edit
6. ^ Scott, EM (2003). "The Fourth

International   Radiocarbon
Intercomparison (FIRI).". *Radiocarbon*
45: 135–285.

7. ^ *a* *b* "NOSAMS Radiocarbon Data and
Calculations".   Woods   Hole
Oceanographic   Institution.
http://www.nosams.whoi.edu/clients/data
.html.

8. ^ Taylor RE, Southon J (2007). "Use of
natural diamonds to monitor ^{14}C AMS
instrument   backgrounds". *Nuclear
Instruments and Methods in Physics
Research   B   **259**:   282–28.
doi:10.1016/j.nimb.2007.01.239.

9. ^ Stuiver M, Reimer PJ, Braziunas TF
(1998). "High-precision radiocarbon age
calibration for terrestrial and marine
samples". *Radiocarbon* **40**: 1127–51.
http://depts.washington.edu/qil/datasets/u
wten98_14c.txt.

10.^ "Atmospheric δ^{14}C record from
Wellington". *Carbon   Dioxide
Information   Analysis   Center*.

http://cdiac.esd.ornl.gov/trends/co2/welli
ng.html. Retrieved 1 May 2008.

11.^ "$\delta^{14}CO_2$ record from Vermunt".
*Carbon Dioxide Information Analysis
Center.*
http://cdiac.esd.ornl.gov/trends/co2/cent-
verm.html. Retrieved 1 May 2008.

12.^ "Radiocarbon dating". Utrecht
University.
http://www1.phys.uu.nl/ams/Radiocarbo
n.htm. Retrieved 1 May 2008.

13.^ Kudela K. and Bobik P. (2004).
"Long-Term Variations of Geomagnetic
Rigidity Cutoffs". *Solar Physics* **224**:
423–431. doi:10.1007/s11207-005-6498-
9.

14.^ Reimer, Paula J.; Brown, Thomas A.;
Reimer, Ron W. (2004). "Discussion:
Reporting and Calibration of Post-Bomb
^{14}C Data". *Radiocarbon* **46** (3): 1299–
1304

15.^ These results were obtained from a
Monte Carlo analysis calibrating

simulated measurements of varying precision using the 1993 version of the calibration curve. The width of the uncertainty represents a 2σ uncertainty (that is, a likelihood of 95% that the date appears between these limits). Niklaus TR, Bonani G, Suter M, Wölfli W (1994). "Systematic investigation of uncertainties in radiocarbon dating due to fluctuations in the calibration curve". *Nuclear Instruments and Methods in Physics Research* **92**: 194–200. doi:10.1016/0168-583X(94)96004-6.

16._^_ Reimer Paula J *et al.* (2004). "INTCAL04 Terrestrial Radiocarbon Age Calibration, 0–26 Cal Kyr BP". *Radiocarbon* **46** (3): 1029–1058. http://digitalcommons.library.arizona.edu/objectviewer? o=http://radiocarbon.library.arizona.edu/Volume46/Number3/azu_radiocarbon_v46_n3_1029_1058_v.pdf. A web interface is here.

17.^ Reimer, P.J.; et. al. (2009). "IntCal09 and Marine09 Radiocarbon Age Calibration Curves, 0–50,000 Years cal BP". *Radiocarbon* **51** (4): 1111–1150. http://researchcommons.waikato.ac.nz/bitstream/10289/3622/1/Hogg%20Intcal09%20and%20Marine09.pdf.

18.^ Balter, Michael (15 Jan 2010). "Radiocarbon Daters Tune Up Their Time Machine". *ScienceNOW Daily News*. http://sciencenow.sciencemag.org/cgi/content/full/2010/115/3.

19.^ Godwin, H. (1962). "Half-life of Radiocarbon". *Nature* **195** (4845): 984. doi:10.1038/195984a0.

20.^ Libby WF (1955). *Radiocarbon dating* (2nd ed.). Chicago: University of Chicago Press.

21.^ Lerman, J. C.; Mook, W. G.; Vogel, J. C.; de Waard, H. (1969). "Carbon-14 in Patagonian Tree Rings". *Science* **165** (3898):   1123–1125.

doi:10.1126/science.165.3898.1123.
PMID 17779805.

22.^ McNichol AP, Schneider RJ, von
Reden KF, Gagnon AR, Elder KL,
NOSAMS, Key RM, Quay PD (October
2000). "Ten years after - The WOCE
AMS radiocarbon program". *Nuclear
Instruments and Methods in Physics
Research, Section B: Beam Interactions
with Materials and Atoms* **172** (1-4):
479–84.   doi:10.1016/S0168-
583X(00)00093-8.

23.^ Stuiver M, Braziunas TF (1993).
"Modelling atmospheric ^{14}C influences
and ^{14}C ages of marine samples to
10,000 BC". *Radiocarbon* **35** (1): 137.

24.^ *a* *b* Kolchin BA, Shez YA (1972).
*Absolute archaeological datings and
their problems*. Moscow: Nauka.

25.^ Crowe C (1958). "Carbon-14 activity
during the past 5000 years". *Nature* **182**
(4633): 470–1. doi:10.1038/182470a0.

26.^ Barker H (1958). "Carbon-14 Activity

during the past 5,000 Years". _Nature_ **182** (4647): 1433. doi:10.1038/1821433a0.

27.^ Libby WF (1962). "Radiocarbon; an atomic clock". _Annual Science and Humanity Journal_.

28.^ Wang YJ; Cheng, H; Edwards, RL; An, ZS; Wu, JY; Shen, CC; Dorale, JA (2001). "A High-Resolution Absolute-Dated Late Pleistocene Monsoon Record from Hulu Cave, China.". _Science_ **294** (5550): 2345–2348. doi:10.1126/science.1064618. PMID 11743199.

29.^ Beck JW; Richards, DA; Edwards, RL; Silverman, BW; Smart, PL; Donahue, DJ; Hererra-Osterheld, S; Burr, GS et al. (2001). "Extremely large variations of atmospheric C-14 concentration during the last glacial period.". _Science_ **292** (5526): 2453–2458. doi:10.1126/science.1056649. PMID 11349137.

30.^ _a_ _b_ Hoffmann DL; Beck, J. Warren;

Richards, David A.; Smart, Peter L.; Singarayer, Joy S.; Ketchmark, Tricia; Hawkesworth, Chris J. (2010). "Towards radiocarbon calibration beyond 28 ka using speleothems from the Bahamas". *Earth and Planetary Science Letters* **289**: 1–10. Bibcode 2010E&PSL.289....1H. doi:10.1016/j.epsl.2009.10.004.

31.^ Jensen MN (2001). "Peering deep into the past". University of Arizona, Department of Physics. http://www.physics.arizona.edu/physics/public/beck-citizen.html.

32.^ Pennicott K (10 May 2001). "Carbon clock could show the wrong time". *PhysicsWeb*. http://physicsworld.com/cws/article/news/2676.

Big Bang Theory

1. ^ D. N. Spergel et al. (2007). "Three-Year Wilkinson Microwave Anisotropy Probe (WMAP) Observations: Implications for Cosmology".

Astrophysical Journal Supplement Series **170** (2): 377–408. arXiv:astro-ph/0603449.   Bibcode 2007ApJS..170..377S. doi:10.1086/513700.

2. ^ *a* *b* Dodelson, Scott (2003). *Modern Cosmology*. Academic Press. ISBN 0-12-219141-2.

3. ^ *a* *b* Liddle, Andrew; David Lyth (2000). *Cosmological Inflation and Large-Scale Structure*. Cambridge. ISBN 0-521-57598-2.

4. ^ *a* *b* Padmanabhan, T. (1993). *Structure formation in the universe*. Cambridge University Press. ISBN 0-521-42486-0.

5. ^ Peebles, P. J. E. (1980). *The Large-Scale Structure of the Universe*. Princeton University Press. ISBN 0-691-08240-5.

6. ^ Kolb, Edward; Michael Turner (1988). *The Early Universe*. Addison-Wesley. ISBN 0-201-11604-9.

7. ^ Wayne Hu and Scott Dodelson (2002).

"Cosmic microwave background anisotropies". *Ann. Rev. Astron. Astrophys.* **40** (1): 171–216. arXiv:astro-ph/0110414. Bibcode 2002ARA&A..40..171H. doi:10.1146/annurev.astro.40.060401.093926.

8. ^ *a* *b* Edmund Bertschinger (1998). "Simulations of structure formation in the universe". *Annual Review of Astronomy and Astrophysics* **36** (1): 599–654. Bibcode 1998ARA&A..36..599B. doi:10.1146/annurev.astro.36.1.599.

9. ^ Harrison, E. R. (1970). "Fluctuations at the threshold of classical cosmology". *Phys. Rev.* **D1**: 2726. Bibcode 1970PhRvD...1.2726H. doi:10.1103/PhysRevD.1.2726.

10.^ Peebles, P. J. E.; Yu, J. T. (1970). "Primeval adiabatic perturbation in an expanding universe". *Astrophysical Journal* **162**: 815. Bibcode 1970ApJ...162..815P.

doi:10.1086/150713.

11.^ Ya; Zel'dovich, B. (1972). "A hypothesis, unifying the structure and entropy of the universe". *Monthly Notices of the Royal Astronomical Society* **160**. Bibcode 1972MNRAS.160P...1Z.

12.^ R. A. Sunyaev, "Fluctuations of the microwave background radiation," in *Large Scale Structure of the Universe* ed. M. S. Longair and J. Einasto, 393. Dordrecht: Reidel 1978.

13.^ U. Seljak and M. Zaldarriaga (1996). "A line-of-sight integration approach to cosmic microwave background anisotropies". *Astrophysics J.* **469**: 437–444. arXiv:astro-ph/9603033. Bibcode 1996ApJ...469..437S. doi:10.1086/177793.

14.^ Springel, V. *et al* (2005). "Simulations of the formation, evolution and clustering of galaxies and quasars". *Nature* **435** (7042): 629–636.

arXiv:astro-ph/0504097.   Bibcode 2005Natur.435..629S. doi:10.1038/nature03597. PMID 15931216.

Quantum Mechanics

1. ^ Richard P. Feynman, *QED*, p. 10
2. ^ Landau, L. D.; E. M. Lifshitz (1996). *Statistical Physics* (3rd Edition Part 1 ed.). Oxford: Butterworth-Heinemann. ISBN 0521653142.
3. ^ This was published (in German) as Planck, Max (1901). "Ueber das Gesetz der Energieverteilung im Normalspectrum". *Ann. Phys.* **309** (3): 553–63. Bibcode 1901AnP...309..553P. doi:10.1002/andp.19013090310. http://www.physik.uni-augsburg.de/annalen/history/historic-papers/1901_309_553-563.pdf . English translation: "On the Law of Distribution of Energy in the Normal Spectrum".
4. ^ Francis Weston Sears (1958).

Mechanics, Wave Motion, and Heat.
Addison-Wesley.   p. 537.
http://books.google.com/books?
hl=en&q=%22Mechanics
%2C+Wave+Motion%2C+and+Heat
%22+%22where+n+%3D+1%2C
%22&btnG=Search+Books.

5. ^ "The Nobel Prize in Physics 1918".
The   Nobel   Foundation.
http://nobelprize.org/nobel_prizes/physic
s/laureates/1918/. Retrieved 2009-08-01.

6. ^ Kragh, Helge (1 December 2000).
"Max   Planck: the   reluctant
revolutionary".   PhysicsWorld.com.
http://physicsworld.com/cws/article/print
/373

7. ^ Einstein, Albert (1905). "Über einen
die Erzeugung und Verwandlung des
Lichtes betreffenden heuristischen
Gesichtspunkt". *Annalen der Physik* **17**:
132–148. Bibcode 1905AnP...322..132E.
doi:10.1002/andp.19053220607.
http://www.zbp.univie.ac.at/dokumente/e

instein1.pdf. , translated into English as On a Heuristic Viewpoint Concerning the Production and Transformation of Light. The term "photon" was introduced in 1926.

8. ^ *a* *b* *c* *d* *e* Taylor, J. R.; Zafiratos, C. D.; Dubson, M. A. (2004). *Modern Physics for Scientists and Engineers*. Prentice Hall. pp. 127–9. ISBN 0135897890.

9. ^ Dicke and Wittke, *Introduction to Quantum Mechanics*, p. 12

10.^ http://ntrs.nasa.gov/archive/nasa/casi.ntrs .nasa.gov/19680009569_1968009569.pd f

11.^ *a* *b* Taylor, J. R.; Zafiratos, C. D.; Dubson, M. A. (2004). *Modern Physics for Scientists and Engineers*. Prentice Hall. pp. 147–8. ISBN 0135897890.

12.^ McEvoy, J. P.; Zarate, O. (2004). *Introducing Quantum Theory*. Totem Books. pp. 70–89, especially p. 89. ISBN 1840465778.

13.^ *World Book Encyclopedia*, page 6, 2007.
14.^ Dicke and Wittke, *Introduction to Quantum Mechanics*, p. 10f.
15.^ J. P. McEvoy and Oscar Zarate (2004). *Introducing Quantum Theory.* Totem Books. p. 110f. ISBN 1-84046-577-8.
16.^ Aezel, Amir D., *Entanglrment*, p. 51f. (Penguin, 2003) ISBN 0-452-28457
17.^ J. P. McEvoy and Oscar Zarate (2004). *Introducing Quantum Theory.* Totem Books. p. 114. ISBN 1-84046-577-8.
18.^ Heisenberg's Nobel Prize citation
19.^ W. Moore, *Schrödinger: Life and Thought*, Cambridge University Press (1989), p. 222.
20.^ Heisenberg first published his work on the uncertainty principle in the leading German physics journal *Zeitschrift für Physik*: Heisenberg, W. (1927). "Über den anschaulichen Inhalt der quantentheoretischen Kinematik und Mechanik". *Z. Phys.* **43** (3–4): 172–198.

Bibcode 1927ZPhy...43..172H.
doi:10.1007/BF01397280.

21.^ Nobel Prize in Physics presentation
speech, 1932

22.^ *a* *b* Linus Pauling, **The Nature of the
Chemical Bond**, p. 47

23.^ E. Schrödinger, *Proceedings of the
Cambridge Philosophical Society*, 31
(1935), p. 555says: "When two systems,
of which we know the states by their
respective representation, enter into a
temporary physical interaction due to
known forces between them and when
after a time of mutual influence the
systems separate again, then they can no
longer be described as before, viz., by
endowing each of them with a
representative of its own. I would not
call that *one* but rather *the* characteristic
trait of quantum mechanics."

24.^ "Quantum Nonlocality and the
Possibility of Superluminal Effects",
John G. Cramer,

http://www.npl.washington.edu/npl/int_r
ep/qm_nl.html

Theory of Relativity

1. ^ Einstein A. (1916 (translation 1920)),
   *Relativity: The Special and General
   Theory*, New York: H. Holt and
   Company
2. ^ Planck, Max (1906), "The
   Measurements of Kaufmann on the
   Deflectability of β-Rays in their
   Importance for the Dynamics of the
   Electrons", *Physikalische Zeitschrift* **7**:
   753–761
3. ^ Miller, Arthur I. (1981), *Albert
   Einstein's special theory of relativity.
   Emergence (1905) and early
   interpretation (1905–1911)*, Reading:
   Addison–Wesley, ISBN 0-201-04679-2
4. ^ *a b c d e f g* Will, Clifford M (August 1,
   2010). "Relativity". *Grolier Multimedia
   Encyclopedia*.
   http://gme.grolier.com/article?

assetid=0244990-0. Retrieved 2010-08-01.

5. ^ _a b c_ Will, Clifford M (August 1, 2010). "Space-Time Continuum". _Grolier Multimedia Encyclopedia._ http://gme.grolier.com/article?assetid=0272730-0. Retrieved 2010-08-01.

6. ^ _a b c_ Will, Clifford M (August 1, 2010). "Fitzgerald-Lorentz contraction". _Grolier Multimedia Encyclopedia._ http://gme.grolier.com/article?assetid=0107090-0. Retrieved 2010-08-01.

7. ^ _a b c d_ Einstein's letter to the London Times in 1919.

   • Einstein Albert (Nov. 28, 1919). "What is the theory of relativity?"". _The London Times_: pp. 4.

ABOUT THE AUTHOR

Don Alexander is 72 years old and lives with his wife, Elaine, in Rocky Mount, Missouri situated along the Lake of the Ozarks. Don earned a triple Bachelor of Science Degree plus a Doctorate in Law from the University of Missouri.

Don also researched physics, chemistry, astrophysics, astronomy, quantum mechanics, biology, genetics, mathematical probabilities, and the Holy Bible over a period of fifty-two years .He has written eight books and two screenplays.

In the Light   Don Alexander